Knowledge Management and Learning

Volume 9

Series Editors

Ettore Bolisani, University of Padua, Padova, Italy

Meliha Handzic, International Burch University, Sarajevo, Bosnia and Herzegovina & Suleyman Sah University, Istanbul, Turkey

This series is introduced by the International Association for Knowledge Management (www.IAKM.net) with an aim to offer advanced peer-reviewed reference books to researchers, practitioners and students in the field of knowledge management in organizations. Both discussions of new theories and advances in the field, as well as reviews of the state-of-the art will be featured regularly. Particularly, the books will be open to these contributions: Reviews of the state-of-the art (i.e. syntheses of recent studies on a topic, classifications and discussions of theories, approaches and methods, etc.) that can both serve as a reference and allow opening new horizons Discussions on new theories and methods of scientific research in organisational knowledge management Critical reviews of empirical evidence and empirical validations of theories Contributions that build a bridge between the various disciplines and fields that converge towards knowledge management (i.e.: computer science, cognitive sciences, economics, other management fields, etc.) and propose the development of a common background of notions, concepts and scientific methods Surveys of new practical methods that can inspire practitioners and researchers in their applications of knowledge management methods in companies and public services.

This is a SCOPUS-indexed book series.

More information about this series at http://www.springer.com/series/11850

Marco Bettiol • Eleonora Di Maria •
Stefano Micelli
Editors

Knowledge Management and Industry 4.0

New Paradigms for Value Creation

Editors
Marco Bettiol
Department of Economics and
Management 'Marco Fanno'
University of Padova
Padova, Italy

Eleonora Di Maria
Department of Economics and
Management 'Marco Fanno'
University of Padova
Padova, Italy

Stefano Micelli
Department of Management
Ca' Foscari University
Venice, Italy

ISSN 2199-8663　　　　　　ISSN 2199-8671　(electronic)
Knowledge Management and Organizational Learning
ISBN 978-3-030-43591-2　　　ISBN 978-3-030-43589-9　(eBook)
https://doi.org/10.1007/978-3-030-43589-9

© Springer Nature Switzerland AG 2020
This work is subject to copyright. All rights are reserved by the Publisher, whether the whole or part of the material is concerned, specifically the rights of translation, reprinting, reuse of illustrations, recitation, broadcasting, reproduction on microfilms or in any other physical way, and transmission or information storage and retrieval, electronic adaptation, computer software, or by similar or dissimilar methodology now known or hereafter developed.
The use of general descriptive names, registered names, trademarks, service marks, etc. in this publication does not imply, even in the absence of a specific statement, that such names are exempt from the relevant protective laws and regulations and therefore free for general use.
The publisher, the authors, and the editors are safe to assume that the advice and information in this book are believed to be true and accurate at the date of publication. Neither the publisher nor the authors or the editors give a warranty, expressed or implied, with respect to the material contained herein or for any errors or omissions that may have been made. The publisher remains neutral with regard to jurisdictional claims in published maps and institutional affiliations.

This Springer imprint is published by the registered company Springer Nature Switzerland AG.
The registered company address is: Gewerbestrasse 11, 6330 Cham, Switzerland

Preface

The topic developed in the book—the relationship between knowledge management and network technologies in the context of firm and its competitiveness—has been at the core of our research interests since the 1990s. Over two decades different technological revolutions took place, where specifically the new economy rooted in the Web has transformed the business landscape forever. Research has been able to identify the opportunities and challenges of the adoption of information and communications technologies (ICT), specifically as far as Small and Medium-sized Enterprises (SMEs) and innovation dynamics are concerned.

In the last five years, digital technologies included under the "Industry 4.0" umbrella concept have opened new potentialities, with a potentially radical shift in the production and distribution of value and business competitiveness across countries. The rise of attention toward such technologies has been particularly emphasized by many national policies oriented to foster the adoption of Industry 4.0 technologies, by stressing the potential gains of the digital manufacturing scenario, organizational transformation, and value chain configurations. The focus on the importance of Industry 4.0 adoption led to underestimation of the implications on how activities are organized within the firm and between firms and in particular how knowledge is produced and shared among organizations. To fill this gap, we tried to reflect on opportunities and threats of the use of Industry 4.0 in a knowledge management perspective.

In this context, the book gathers theoretical insights and empirical analyses related to the research project carried between 2017 and 2019 at the University of Padova (Department of Economics and Management) and focused on the digital transformation of manufacturing firms (project "Manufacturing activities and value creation: redesigning firm's competitiveness through digital manufacturing in a circular economy framework"). Contributions to the book refer to scholars who presented and discussed their research in relation to the final Workshop on "Creating Value Through Manufacturing: Exploiting Industry 4.0 in a Circular Economy Framework" held in Padova on March 14–15, 2019. This workshop gathered Italian and German scholars to compare and debate their studies.

In addition, further theoretical refinements included in the book are rooted in the experience of the Venice International University—VIU Graduate Seminar "Rethinking manufacturing, consumption, and globalization in the era of automation" held at the Venice International University, Venice, on September 9–13, 2019. The VIU Graduate Seminar has been proposed to coincide with the 20th anniversary of the TeDIS Center of VIU (now TeDIS program) launched in 1999.

Twenty years ago, the TeDIS Center has been developed proposing a specific line of research devoted to study the relationship between the rise of network technologies and the model of organization of economic activities. In this new digital context, the VIU Graduate Seminar has provided the opportunities to theoretically deepening the understanding of the challenges and chances of this Industry 4.0 scenario, suggesting a framework to interpret the transformations that will occur at multiple levels in manufacturing and consumption. Insights emerged from the VIU Graduate Seminar have been included in the Introduction and became inputs for the chapters of the book.

The editors are grateful to the authors for their rich analyses and fresh views provided on the relevant topic explored in the book and to the extraordinary experience lived throughout those 20 years within the TeDIS—VIU.

Padova, Italy	Marco Bettiol
Padova, Italy	Eleonora Di Maria
Venice, Italy	Stefano Micelli

Acknowledgments

This publication has been supported by the project "Manufacturing activities and value creation: redesigning firm's competitiveness through digital manufacturing in a circular economy framework"—grant no. BIRD161248/16 (University of Padova, Department of Economics and Management) (further information available at: https://www.economia.unipd.it/en/DML/Digital-Manufacturing-Lab).

The editors would like to thank the participants of the Workshop "*Creating Value Through Manufacturing: Exploiting Industry 4.0 in a Circular Economy Framework*" held in Padova on March 14–15, 2019, and of the VIU Graduate Seminar "*Rethinking manufacturing, consumption, and globalization in the era of automation*" held at the Venice International University, Venice, on September 9–13, 2019.

Contents

Industry 4.0 and Knowledge Management: An Introduction 1
Marco Bettiol, Eleonora Di Maria, and Stefano Micelli

Industry 4.0 and Knowledge Management: A Review of Empirical Studies 19
Mauro Capestro and Steffen Kinkel

Do Industry 4.0 Technologies Matter When Companies Backshore Manufacturing Activities? An Explorative Study Comparing Europe and the US 53
Luciano Fratocchi and Cristina Di Stefano

Knowledge and Digital Strategies in Manufacturing Firms: The Experience of Top Performers 85
Marco Bettiol, Mauro Capestro, Eleonora Di Maria, and Stefano Micelli

Industry 4.0 and Creative Industries: Exploring the Relationship Between Innovative Knowledge Management Practices and Performance of Innovative Startups in Italy 113
Silvia Blasi and Silvia Rita Sedita

Coordinating Knowledge Creation: A Systematic Literature Review on the Interplay Between Operational Excellence and Industry 4.0 Technologies 137
Toloue Miandar, Ambra Galeazzo, and Andrea Furlan

Achieving Circular Economy Via the Adoption of Industry 4.0 Technologies: A Knowledge Management Perspective 163
Valentina De Marchi and Eleonora Di Maria

Industry 4.0: New Paradigms of Value Creation for the Steel Sector ... 179
Laura Tolettini and Claudia Lehmann

List of Contributors

Marco Bettiol Department of Economics and Management 'Marco Fanno', University of Padova, Padova, Italy

Silvia Blasi Department of Economics and Management 'Marco Fanno', University of Padova, Padova, Italy

Mauro Capestro Department of Economics and Management 'Marco Fanno', University of Padova, Padova, Italy

Valentina De Marchi Department of Economics and Management 'Marco Fanno', University of Padova, Padova, Italy

Eleonora Di Maria Department of Economics and Management 'Marco Fanno', University of Padova, Padova, Italy

Cristina Di Stefano Department of Industrial and Information Engineering and Economics, University of L'Aquila, L'Aquila, Italy

Luciano Fratocchi Department of Industrial and Information Engineering and Economics, University of L'Aquila, L'Aquila, Italy

Andrea Furlan Department of Economics and Management 'Marco Fanno', University of Padova, Padova, Italy

Ambra Galeazzo Department of Economics and Management 'Marco Fanno', University of Padova, Padova, Italy

Steffen Kinkel University of Applied Sciences, Karlsruhe, Germany

Claudia Lehmann HHL Leipzig Graduate School of Management, Leipzig, Germany

Toloue Miandar Department of Economics and Management 'Marco Fanno', University of Padova, Padova, Italy

Stefano Micelli Department of Management, Ca' Foscari University, Venice, Italy

Silvia Rita Sedita Department of Economics and Management 'Marco Fanno', University of Padova, Padova, Italy

Laura Tolettini Feralpi Holding SpA, Lonato del Garda, Italy

Industry 4.0 and Knowledge Management: An Introduction

Marco Bettiol, Eleonora Di Maria, and Stefano Micelli

Abstract The fourth industrial revolution promises to deeply transform the business landscape offering new opportunities to create and use knowledge. However, firm's knowledge management strategies have been supported by technological investments for decades. The chapter explores two prior "revolutions" connected to the digital technologies—ERP and the Web—and their implications for knowledge management dynamics, identifying how Industry 4.0 technologies can further enhance those processes and the related challenges. The main contributions from the book are outlined in terms of relationships between Industry 4.0 technologies and competences and geographical implications, focusing on new firms and connection with the two strategic goals of operational excellence and environmental sustainability.

1 Introduction

The study related to knowledge management and the firm is vast and has covered a large set of topics. However, it is far from being exhaustive since the emerging dynamic technological scenario related to digital technologies in general and Industry 4.0 ones specifically asks for further attention on how knowledge is created, by whom, for what purpose, and with which outcomes.

In management studies a large set of contributions have referred to knowledge management as the key concept and relevant process firms have to deal with. From a strategic management perspective, firms manage knowledge in order to sustain their competitiveness and knowledge becomes the driver for the firm's competitive

---

M. Bettiol · E. Di Maria (✉)
Department of Economics and Management 'Marco Fanno', University of Padova, Padova, Italy
e-mail: eleonora.dimaria@unipd.it

S. Micelli
Department of Management, Ca' Foscari University, Venice, Italy

advantage (Kogut & Zander, 1996). Differently from other theoretical frameworks such as the transaction cost theory (Coase, 1937), the firm as a decision-making mechanism (Cyert & March, 1963), or the agency theory of the firm (Alchian & Demsetz, 1972), according to the knowledge-based theory of the firm, the firm can be interpreted as a mechanism oriented to managing knowledge (Nonaka, 2000; Spender, 1996). Knowledge becomes a key resource for firms, which have to transform individual knowledge into organizational knowledge (Grant, 1996) through appropriate processes of organizational learning.

Creating, elaborating, and transferring knowledge define different steps of the knowledge management process and imply a variety of actors involved—both within and outside the firms taking into consideration the variety of forms of knowledge (Nonaka & Takeuchi, 1995). Knowledge management is seen as an open, distributed, and extended process where individuals and organizations are connected (Argyris & Schon, 1978; March, 1991; Pfeffer & Sutton, 2000). This perspective has guided studies on innovation, which consider both technological and user-driven forms of innovation and the degree of openness of such dynamics (Chesbrough, 2003; Jensen, Johnson, Lorenz, & Lundvall, 2007). From a knowledge sourcing perspective, actors within the firms with different levels of specialization and background are relevant (from R&D to marketing, from managers to blue-collar employees), but also actors external to the organizational boundaries are as relevant as internal ones—KIBS, suppliers, and customers (Di Bernardo, Grandinetti, & Di Maria, 2012; Laursen & Salter, 2006). This is also reflected in the geography of knowledge management, since knowledge sources can benefit not only from physical proximity but also from other forms of proximity, such as social, relational, or cognitive ones (Bathelt, Malmberg, & Maskell, 2004; Boschma, 2005; Brown & Duguid, 2001).

In terms of processes of knowledge management, much attention has been given to the level of knowledge codification to evaluate the degree of "stickiness" of knowledge transfer (Szulanski, 2000; von Hippel, 1994) and how knowledge can flow across actors and places (Cowan, David, & Foray, 2000). With the idea of exploring the transformation of tacit knowledge into codified one, many studies have put the attention on the role of information systems and on information and communication technologies more in general. Among the many studies, Hansen, Nohria, and Tierney (1999) provide a relevant contribution to the debate by suggesting two possible strategies to managing knowledge transfer among individuals (organizations) also across the space: the people-to-people and people-to-document strategies. While in the first one the emphasis is on personal interaction and social dynamics to allow knowledge flows—especially in case of complexity of the knowledge to be transferred—the second one is instead more focused on the exploitation of information technological solutions—digital knowledge repositories—to collect embodied knowledge, allowing distributed access. In the knowledge processes of knowledge creation and acquisition/transfer different technological tools can be used (Baskerville & Dulipovici, 2006). The need to put attention toward the human side of application of information technologies for effective knowledge management purposes has been stressed also by Davenport (1994), who was among the many scholars debating this topic. Davenport explores information technologies for knowledge management quite extensively (Davenport, 2007; Davenport & Prusak, 2000). According to his study, different types of complexity of the work to be done by knowledge workers—routine vs. expert model/judgement/interpretation—as well as the level of interdependency (individual vs. collaborative groups) generate four different types of knowledge works which are associated with different sets of information technologies that can support such works. Alternative technological solutions should fit with the variety of knowledge to be managed: the more open, collaborative, and complex is the work, the more is required for firms to have knowledge repositories and collaborative tools, relying also on data mining and analytics solutions. On the contrary, process applications and workflow management or transactional technologies are more consistent with routinized, individual works, while decision automation is for medium complex knowledge work. The rise of information systems as technological systems oriented to support organization process can be interpreted as a revolution occurring at the knowledge management level. New set of technologies not focused on manufacturing—in the operations department—but oriented to enhance the work of knowledge workers (in the office) increase their productivity (Drucker, 1999).

This scenario has been further enlarged with the rise of the Web and the opportunity for other actors to be involved in the knowledge management process outside the organization (Kozinets, Hemetsberger, & Schau, 2008), where the role of virtual communities has been emphasized (Rheingold, 1993). From this point of view, the new economy has been seen as another technology-driven revolution, where the focus is on the C in the information and communications technologies (ICT) context. Beyond data management, the Web became the digital tool stressing the connectivity potentiality and the transformation of forms of interaction at distance among actors—especially in the consumer sphere (Armstrong & Hagel, 1996). Within the studies on lead users and customer-centric innovation processes (von Hippel, 1986), the Web became the enabling infrastructure to provide customers new toolkits (Von Hippel, 2001) and the new digital environment when distributed innovation could take place (Nambisan & Nambisan, 2008).

Now the rise of new technologies related to Industry 4.0 (Schneider, 2018; Ustundag & Cevikcan, 2018) is further changing the knowledge management framework and how firms can leverage on such technologies for managing knowledge and enhance their competitive advantage. Industry 4.0 as fourth industrial revolution is the new competitive context in which the firm defines its strategy and through which it creates value (Reinhard, Jesper, & Stefan, 2016). Industry 4.0 challenges strategic processes in terms of value generated within different activities of the value chain (value of manufacturing) (Rehnberg & Ponte, 2018) as well as geography (Buunge & Zucchella, 2017).

From a knowledge management standpoint, Industry 4.0 opens new questions on the process of organizational learning (knowledge creation, codification, and transfer)—within the firm (Nonaka & Takeuchi, 1995)—but also across actors—may they be firms (networks) (Inkpen, 1996), but also customers/users (communities) (Brown & Duguid, 2001). The book aims at presenting the challenges of managing knowledge in the context of Industry 4.0. The development of digital technologies applied to manufacturing (additive manufacturing, IoT, robotics, etc.) suggests a paradigmatic change in value creation. Through theoretical and empirical contributions, the book provides insights on the way the Industry 4.0 technologies allow firms to create and exploit knowledge. New technologies offer the opportunity to acquire knowledge from a larger number of sources and open issues on how firms approach innovation, organize activities, and develop new relationships with their stakeholders in order to deliver on the market customized products and services.

2 Knowledge Management and Technological Revolutions

2.1 Knowledge Management, BPR, and ERP

The first managerial revolution linked to the introduction of digital in companies coincides with the introduction of so-called integrated management systems as a digital infrastructure for the management of companies. These technologies represented an important technological leap forward compared to the traditional IT tools available to companies since the end of the 1980s. Up to that date, business IT had developed software solutions capable of responding to specific business functions (administration, finance, production, marketing, etc.) without it being possible to rely on shared databases and without being able to rely on simple solutions for the management of inter-functional processes. Software for integrated business management (i.e., the world leader SAP) named Enterprise Resource Planning (ERP) overturned this development logic by offering a single platform with different applications for different business functions and a single reference database, transforming IT in a fundamental driver for value creation rooted in new intelligence built in the software (Gable, Scott, & Davenport, 1998; Micelli, 2017; Nevo & Wade, 2010).

The introduction of these integrated management solutions triggers a profound change in the functioning of the organizational dynamics of large multinational companies, through a deep transformation called Business Process Reengineering (BPR). This transformation coincides with a substantial change in the way organizations are managed, going beyond the functional organization to adopt a model structured by processes thanks to technology (Hammer & Champy, 2001). This literature explicitly highlights the radical dimension of change imposed by new technologies and proposes methods of intervention that aim to implement radically new forms of process organization.

Information technology (IT) supports coordination within and among organizational units, where the BPR logic transforms the firm's approach toward activities by overcoming function boundaries to stress the interconnectivity among them. IT becomes relevant in the BPR during different phases of BPR (Attaran, 2004): in the design phase, technologies enable collecting, codification, and transfer of knowledge related to the different activities, actors, and locations involved in the process to be redesigned; in the implementation phase, IT allows gathering and analyzing the information concerning the performance and structure of the processes, where codification of the process is intertwined with the codification of the knowledge related to it; after BPR is implemented, IT monitors and sustains coordination also at distance.

ERP represents the technological dimension of the firm processes and its implementation is connected with the adoption of a BPR logic within the firm. "The key underlying idea of ERP is to achieve a capability of planning and integrating enterprise-wide resources" (Xu, Wang, Luo, & Shi, 2006, p. 148). Especially in geographically dispersed enterprises, ERP has been used for knowledge management purposes, by coordinating the connection between the core and the periphery of actors in the network (Lipparini, Lorenzoni, & Ferriani, 2014). ERP provides a large amount of data and information related to multiple processes and actors through a unified view, thus becoming a channel for knowledge management strategies. At the same time, clear knowledge management strategies the firm aims at developing have also an impact on ERP adoption and functioning within the organization (Xu et al., 2006). Within the ERP framework, much attention has been given to knowledge codification within the company. Tacit knowledge challenges ERP implementation, pushing firms to consider also the relevance of individuals within the organization. Codification and socialization approaches have to be coupled to benefit more from this intersection between knowledge management and IT systems (Apostolou, Abecker, & Mentzas, 2007), taking into account that knowledge creation and transfer can rely on sharing the same practice among individuals of the same organization (communities of practice) (Brown & Duguid, 2000).

2.2 Knowledge Management and the Web

A second important revolution in the application of digital technologies in firms is related to the Web. Since the end of the 1990s, the spread of the Internet and the changes that Internet would have triggered were thought of as the beginning of a real "new economy," with its own rules, specific and different from those that have governed the economy of the past (Kelly, 1998; Porter, 2001). There is little doubt that the rapid spread of the Web has constituted a substantial revolution in the way of doing business, particularly with regard to the relationship between business and consumer. The latter, far from being alone and isolated in its decision-making processes, can leverage digital social connections allowing a previously unknown capacity for evaluation and proposal. This knowledge contribution represents an opportunity for companies to grow and innovate (Armstrong & Hagel, 1996). At the same time, the Web strongly impacts on the traditional distribution channels, toward a larger variety of e-commerce strategies (Gulati & Garino, 2000).

A growing body of studies have explored the involvement of customers into the firm's innovation processes. Starting from the seminal contribution by von Hippel (Von Hippel, 1978), scholars stress how lead users and customers may bring their knowledge into the dynamics of product development internal to the firm and more in general into the firm's activities (Cova & Dalli, 2009; Gruner & Homburg, 2000). Digital technologies and the rise of distributed, open, communication infrastructures reduce barriers and costs for customer participation (Bagozzi & Dholakia, 2006; Galvagno & Dalli, 2014). With respect to the mass production paradigm, the Web opens new trajectories in the relationship between the firm and its customers, shifting the power from the firm to the customers. More importantly, customers (users) define different forms of aggregation—the communities—which can also become autonomous actors in the process of innovation and knowledge management (Baldwin, Hienerth, & von Hippel, 2006; Boudreau & Lakhani, 2009; Di Maria & Finotto, 2008; Sawhney, Verona, & Prandelli, 2005).

In order to create a more stable and interactive relationships with customers, the firm may develop specific toolkits for innovation (Franke & Piller, 2004; Von Hippel, 2001). Through the Web, the firm creates virtual spaces to exchange knowledge with its online communities (Armstrong & Hagel, 1996; Füller, Jawecki, & Mühlbacher, 2006; Kozinets, 1999). Specifically, firms with strong brand invest in order to gather knowledge from specialized customers and lead users (Marchi, Giachetti, & De Gennaro, 2011) within a new scenario of co-creation also of the intangible dimension of the brand perceived as shared asset with customers (Fueller & Von Hippel, 2008; Schau, Muñiz, & Arnould, 2009). Virtual Customer Environments become the digital channels for knowledge management (Nambisan & Nambisan, 2008): customers are involved in new product development through idea generation; in product testing through collection of feedbacks; and in product marketing through electronic supports to other customers based on the individual experience and knowledge.

The business scenarios generated by ICT show a transformation in firm's internal processes as well as in the connection with the market. In this perspective, they have prepared the basis for the new revolution related to Industry 4.0.

3 Opportunities and Challenges for Knowledge Management in the Industry 4.0 Context

The "Industry 4.0" label includes a large variety of technologies with different characteristics and domains of applications (Reinhard et al., 2016). Connected robotics, advanced automation, and sensors may transform the firm and specifically operations, driving toward the rise of a smart factory (Büchi, Cugno, & Castagnoli, 2020; Mittal, Khan, Romero, & Wuest, 2018). Not only manufacturing processes can be enhanced in terms of efficiency, but also new extended and detailed control opportunities may rise. A challenge is connected to the exploration of relationship between the implementation of Industry 4.0 technologies and lean management strategies (Kamble, Gunasekaran, & Dhone, 2020; Sanders, Elangeswaran, & Wulfsberg, 2016), where lean management has been one of the key new managerial practices developed during the 1990s that strongly impacted on manufacturing organization (and beyond) (Womack & Jones, 1997). In this perspective, learning processes occurring in relation to lean and operational excellence practices may benefit from digitalization and the implementation of Industry 4.0 technologies. At the same time, however, because of lean experience the firm may implement and exploit such technologies better (Rauch, Dallasega, & Matt, 2016).

Moreover, a growing body of research is exploring the consequences in terms of job metamorphosis connected to automation (and also artificial intelligence; see below) (Autor, 2015; World Economic Forum, 2016). On the one hand, studies highlight the reorganization of manufacturing activities toward the decrease in jobs in operations, while on the other hand research explores also the opportunity of skill redesign due to new forms of interaction and collaboration with technologies (Bakhshi, Downing, Osborne, & Schneider, 2017). In terms of knowledge management, this opens challenges in the types of knowledge firms and workers have to invest in: from substitution of jobs in more routinized activities toward new roles that workers can play going beyond specialization to include a more interdisciplinary approach (Pfeiffer, 2016). In this perspective, scholars are also investigating the geographical implications for manufacturing location related to Industry 4.0 (Dachs, Kinkel, & Jäger, 2019; Rehnberg & Ponte, 2018; Strange & Zucchella, 2017).

This transformation in the factories is also more and more connected to the environmental sustainability side of production, where circular economy framework calls for a better understanding and measurement in the use of resources (Tseng, Tan, Chiu, Chien, & Chi, 2018). The factory is not only smart, but through such technologies may also become green (Bonilla, Silva, Terra Da Silva, Gonçalves, & Sacomano, 2018). Scholars and practitioners are exploring how to use digital technologies as a means of achieving better environmental goals, in particular within the circular economy framework (Lacy & Rutqvist, 2016; Webster & MacArthur, 2017). The need for new circular-oriented innovation and the collaborative dimension of eco-innovation (Brown, Bocken, & Balkenende, 2019; De Marchi, 2012) push firms to exchange knowledge within the value chain—both upstream and downstream.

Industry 4.0 technologies open interesting opportunities of controlling use of resources and sustain knowledge exchange with the actors involved (Tseng et al., 2018).

This issue is related to studies on 3D printing. On the one hand, this technology is linked to the rise of the new paradigm of mass customization, where the firm can mix variety of products with efficiency in small-scale production also with the involvement of customers (Bogers, Hadar, & Bilberg, 2016). Such approach is considered a form of direct digital manufacturing, involving customers in the production (Holmström, Holweg, Khajavi, & Partanen, 2016). The maker movement is becoming protagonist of this revolution (Anderson, 2012), where customers have a new tool—3D printing—they can use not only to design, but actually to physically produce the product (Kalva, 2015; Laplume, Anzalone, & Pearce, 2016). On the other hand, 3D printing is considered a driver for new business models, also in relation to the circular economy framework (Despeisse et al., 2017; Unruh, 2019).

In addition to the above mentioned technologies, scholars and managers devoted much attention to big data and artificial intelligence. Such technologies can revolutionize the way through which firms collect and manage data, but also control learning processes and develop business scenarios (Boden, 2016; Kaplan & Haenlein, 2019; Ransbotham, Khodabandeh, Fehling, LaFountain, & Kiron, 2019). The debate concerning the relationship between knowledge management and artificial intelligence (AI) is not new (Wiig, 1999). However, compared to previous forms of AI, in the present scenario such cognitive processes are not only automated but also augmented such as the definition of Analytics 4.0 (Davenport, 2018a).

From a knowledge management perspective, the first opportunity connected to Industry 4.0 technologies refers to the fact that new processes and tools are available to acquire and elaborate distributed knowledge. Through IoT the firm is able to obtain a growing, constant (ongoing), detailed, customized set of data both from internal sources (i.e., smart machines enacted by sensors) and from external ones (smart products used by customers) (McKinsey Global Institute, 2015). Such data integrated within ERP and more in general the firm information systems can be analyzed through advanced processes of data analysis related to big data analytics and AI (Liebowitz, 2001). In this perspective, the firm may know more about its processes and products from multiple perspectives—marketing, innovation, operations, administration (i.e., Paschen, Kietzmann, & Kietzmann, 2019).

A big challenge refers to the ability of the firm to translate data into knowledge (Pauleen, 2017; Tian, 2017): it is not an automatic process the learning dynamics that an organization can develop through big data. In order to obtain answers from problem-solving situation and take decisions and actions, the firm applies analytics to big data databases to extract information and identify new knowledge also based on—and in coordination with—contextual knowledge inputs (Pauleen & Wang, 2017). Through AI and big data analysis, the firm could be able to augment its strategic vision as it may rely on new resources to strengthen its competitive advantage (Davenport, 2018b). As in the case of automation (robotics), also for AI new open questions refer to the kind of competences required within the firm to exploit such technologies, but also the implications in terms of job transformation (Daugherty & Wilson, 2018).

The second opportunity is linked to the actors that can be involved in the knowledge management dynamics. New actors are producing knowledge within open (autonomous) innovation processes. As mentioned, makers are customers that exploit 3D printing to create and produce new products, customized on their desires and needs, independently or through makerspaces (Halbinger, 2018; Kohtala & Hyysalo, 2015). From this point of view, firms able to connect to those customers for innovation purposes could benefit from their knowledge from a user-driven perspective, beyond the digital tools offered by the Web (Von Hippel, 2001). With the rise of big data and AI a new question arises and it is related to the level of exploitation of data generated by customers. More generally, scholars suggest the positive impacts of Industry 4.0 technologies in the customer relationship management: through smart products (Porter & Heppelmann, 2014) and digital ubiquity

(Iansiti & Lakhani, 2014) the firm may gather more fine-grained knowledge on the use of the product constantly. This approach asks for explicit knowledge management strategies to be adopted in the new digital scenario, since it is still not clear to what extent additional large amount of new knowledge is really relevant or how to defend the firm's competitive advantage on it (Hagiu & Wright, 2020).

A new actor that should be included among the knowledge sources are the machines themselves. As claimed by studies on artificial intelligence (Boden, 2016), machines can be seen as new digital agents who may act independently from the human inputs. According to Floridi "Artificial agents are not science fiction but advanced transition systems capable of interactive, autonomous and adaptable behavior" (Floridi, 2005, p. 416). Through AI, machine learning processes can generate additional knowledge as inputs for strategic developments. Within the theoretical debate on AI (and IT more in general) and knowledge management (Liao, 2003), the novelty of the present scenario refers to the availability of big data, improved computational power, and a system of interconnected technologies that enhance the production of "autonomous" knowledge to be used at the firm level (Yao, Zhou, Zhang, & Boer, 2017; Zhuang, Wu, Chen, & Pan, 2017). This scenario is connected to the problem of control and validation of the new knowledge created through AI outputs with respect to emerging processes such as AI-driven design processes (generative design). By combining multiple perspectives and tools, knowledge-based engineering improves product development through the autonomous inputs of technology, where the focus is on repetitive and non-creative design tasks but also to support multidisciplinary design optimization (Rocca, 2012). Even if the relationship between AI and creativity (and design) is not new (Boden, 1998), new approaches in product design are connected to the opportunity of sustaining the competitiveness related to mass customization within the Industry 4.0 framework (Zawadzki & Żywicki, 2016).

It emerges also a potentially reduced problem of exploration (March, 1991) in the Industry 4.0 scenario, where technological solutions connecting big data and advanced analytics increase the efficiency in gathering data and potentially transform them into knowledge—also overcoming the geographical limitations. However, also issues related to the control over such knowledge emerge. The rise of platform economy (Parker, Van Alstyne, & Choudary, 2016) and the disproportion of power between large firms and SMEs (Wu & Gereffi, 2018) may reduce especially for the latter opportunities of knowledge creation and exploitation. At the same time, firms with prior IT investments and experience in previous revolutions—the EPR/BPR and Web ones mentioned above—can benefit more from the fourth industrial revolution, with respect to firms that have neither clear digital strategy nor past experience on how to effectively introduce technologies within the organization, irrespective of the size. It is not a matter of investing in all the Industry 4.0 technologies available simultaneously (Reinhard et al., 2016), rather to choose the right, appropriate technologies for the product and the processes that characterize the firm and its strategy (Bettiol, Capestro, Di Maria, & Furlan, 2019; McAfee, 2004).

3.1 Managing Knowledge Within the Industry 4.0 Scenario: Contributions of the Book

In line with the theoretical picture depicted above, the book aims at exploring the relationship between Industry 4.0 technologies and knowledge management dynamics from different perspectives.

The contribution by Capestro and Kinkel (Chapter "Industry 4.0 and Knowledge Management: A Review of Empirical Studies") provides insights of the linkages between the fourth industrial revolution and knowledge management issues through a literature review aiming at identifying how scholars have explored empirically this topic. The authors analyze about 50 empirical studies focused on the implementation of a set of Industry 4.0 technologies to explore their relationships with knowledge management. They emphasize the knowledge implications related to processes, products, and people, suggesting the need for manufacturing firms to develop clear strategic orientation toward knowledge management. Moreover, firms should invest in order to upgrade human resources' competences to also include digital skills.

The analysis on the implications of Industry 4.0 technologies on value chain activities and their geographical location has been explored by Fratocchi and Di Stefano (Chapter "Do Industry 4.0 Technologies Matter When Companies Backshore Manufacturing Activities? An Explorative Study Comparing Europe and the US"). A growing number of studies—especially recently with the rise of attention and the diffusion of Industry 4.0 technologies—are evaluating the transformation in the location choices of manufacturing activities of firms from advanced countries. The theoretical premises suggest the cost advantages of the new technological landscape, pushing manufacturing firms in redesigning their offshoring strategies across countries, also considering the home country as an option (compared to the past decades). Fratocchi and Di Stefano developed an extensive structured literature review followed by an analysis of empirical evidence of backshoring decisions implemented by both European and US firms. The authors suggested that even if theoretically Industry 4.0 technologies have been identified as drivers for backshoring decisions, the empirical analysis identifies those technologies both as driver and as enabling factor. Among the many technologies, only automation (robotics) and 3D printing are cited as relevant in backshoring decisions. Moreover, there are differences in the way European vs. US firms adopt such technologies. Most importantly, it seems that automation per se is not necessarily related to decisions concerning backshoring from low-cost countries to high-cost country. This result is explained in terms of competences available and developed within firms adopting Industry 4.0 technologies. It is not a matter of technological investments, rather of coupling competences in the use of Industry 4.0 technologies with knowledge related to manufacturing processes. From this point of view, the adopting firm should develop appropriate learning dynamics in order to effectively exploit the advantages of Industry 4.0 technologies within its backshoring strategies.

In their analysis of knowledge management strategies in top performers, Bettiol et al. (Chapter "Knowledge and Digital Strategies in Manufacturing Firms: The

Experience of Top Performers") suggest that Industry 4.0 technologies may have different impacts in terms of how knowledge is created and shared at the firm level. Compared to previous scenarios, the emerging new technological one has potential implications on the knowledge related to products, but also on the knowledge related to Industry 4.0 technologies themselves (new potentialities that have to be fully discovered in their synergies yet) as well as knowledge of the management of the firm (learning on how to exploit Industry 4.0 technologies, also in connection with prior ICT revolutions). On the one hand advanced, interconnected technologies generate new knowledge autonomously, but on the other hand, in order to really deploy the value connected to data produced by such technologies, firm should also rely on the social dimension of knowledge management dynamics. The empirical analysis on *Champions* shows how specifically AI is intertwined with other data-driven technologies—cloud, big data, and IoT—in addition to prior ICT investments, where skills and competences related to Industry 4.0 are critical in order to manage effectively the implementation and use of digital technologies.

Blasi and Sedita study the implications on knowledge management and innovation of adopting Industry 4.0 technologies in firms specializing in the creative industries. In such industries knowledge management is characterized by the relevance of interaction between the firm and its suppliers and customers, in relation to business innovation processes. The focus of the empirical analysis in Chapter "Industries 4.0 and Creative Industries: Exploring the Relationship Between Innovative Knowledge Management Practices and Performance of Innovative Startups in Italy" is Italian startups in creative industries in general and ICT specifically. Within the Industry 4.0 scenario, startups not only can benefit from being early adopters but also can play a role as promoters of Industry 4.0 technologies in their markets (as creative industries). Three clusters of firms emerged with respect to the adoption of Industry 4.0 technologies (taking into account specifically IoT and big data) as well as their turnover: smart adopters, regular adopters, and laggards. Through a deep analysis of factors characterizing the three clusters—specifically as far as funding sources are concerned—the authors suggest that adopting Industry 4.0 technologies can be an opportunity, but also a challenge in particular for the first two clusters of startups. In fact, strong innovators (Industry 4.0-wise) have different funding forms of sourcing and a reduced portfolio of resources with respect to laggards. Such results highlight the need to further explore the consequences of innovation (and knowledge management) in advanced technologies for startups and their sustainability over time.

The interrelation between innovation and Industry 4.0 technologies can be explored also by examining organizational innovation and specifically the redesign of production processes connected to the achievement of operational excellence. With respect to startups described above, in this perspective established manufacturing firms may decide to adopt Industry 4.0 technologies to obtain benefits of efficiency and better control on operations, considering the opportunities connected to the smart factory. However, such relationship is far from being exhaustive in its implementation and it is the content of the contribution by Miandar, Galeazzo, and Furlan (Chapter "Coordinating Knowledge Creation: A Systematic Literature

Review on the Interplay Between Operational Excellence and Industry 4.0 Technologies"). The authors develop a systematic literature review in order to evaluate the interplay between these two new sources of knowledge. Four different paths emerge taking into account the direction of relationships between Industry 4.0 and operational excellence. Most of the studies stress that such technologies are enablers of lean manufacturing, where firms can exploit the technological potentialities to obtain results of operational excellence. By considering the knowledge management framework proposed by Nonaka and the Thompson's inputs on task coordination, the results of the analysis suggest that Industry 4.0 and operational excellence should be coordinated sequentially. Moreover, not all Industry 4.0 technologies support those dynamics. Limited evidence is related instead of other forms of connections between Industry 4.0 technologies and operational excellence.

The last two chapters of the book study Industry 4.0 technologies in relation to environmental sustainability strategies of firms. In Chapter "Achieving Circular Economy via the Adoption of Industry 4.0 Technologies: A Knowledge Management Perspective", De Marchi and Di Maria show the presence of relevant differences between green and non-green adopters when Industry 4.0 technologies and sustainability are concerned. Through an empirical analysis of Italian manufacturing firms, the authors analyze how the adoption of Industry 4.0 technologies supports the achievement of green outcomes, with special attention to the circular economy framework. It emerges that those technologies support manufacturing adopters in obtaining sustainability outcomes, in case of both proactive, circular-oriented firms and firms that discover sustainability as a non-intended consequence of technological investments. In general, green adopters have a higher investment rate of Industry 4.0 technologies. Moreover, despite the theoretical expectations on 3D printing, specifically robotics and augmented reality are the technologies where there is a statistically significant difference between green and non-green adopters. The focus is specifically on manufacturing activities as domain of investments—confirming the strong connection between circular economy and production and the importance of Industry 4.0 technologies in this relationship. From a knowledge management point of view, technologies help firms in supporting the achievement of sustainability outcomes based on collaboration within the firm boundaries (among workers and functions) as well as in the value chains.

In the conclusive Chapter, Tolettini and Lehmann grounded such theoretical discourse within the steel sector (Chapter "Industry 4.0: New Paradigms of Value Creation for the Steel Sector") in Germany and Italy. Through a deep and extensive study of the industry, the authors focus on the Feralpi case study to explain the strategic potentialities of Industry 4.0 technologies in renovating competitiveness of firms in the steel sector. With particular emphasis on environmental (and social) sustainability, the authors show the advantages steel firms may achieve through technological investments in terms of resource efficiency and product and process innovation, up to more flexible supply chain and more reactive market response. The case study of Feralpi—as innovative leading company in the industry at the national and international level—suggests that many Industry 4.0 technologies can be adopted by a steel company in multiple steps of the value chain and for multiple goals. In particular, technological investments increase the ability of the firm in knowing operations better (especially in most critical phases) through data-driven approach, allowing the firm in reducing risks and negative outputs for workers and the local community among the many stakeholders considered. Moreover, Industry 4.0 technologies support strategic decisions for firm's evolution within the industry.

To conclude, theoretical and empirical studies included in the book further advance knowledge on knowledge management implications of the adoption of Industry 4.0 technologies. Our book suggests that there is a strong impact of such technologies in the processes of knowledge creation and transfer, but also that the fourth industrial revolution should be interpreted as a long-term transformation, where the implementation of technologies itself generates learning processes which do not lead only to immediate results. Further research on this issue is also required in the forthcoming years to have a more complete picture of the strategic consequences of Industry 4.0.

References

Anderson, C. (2012). *Makers: The new industrial revolution.* New York: Crown Business Books. https://doi.org/10.1016/0016-3287(91)90079-H.

Alchian, A. A., & Demsetz, H. (1972). Production, information costs, and economic organization. *The American Economic Review, 62,* 777–795.

Apostolou, D., Abecker, A., & Mentzas, G. (2007). Harmonising codification and socialisation in knowledge management. *Knowledge Management Research and Practice, 5*(4), 271–285. https://doi.org/10.1057/palgrave.kmrp.8500156.

Argyris, C., & Schon, D. A. (1978). *Organizational learning: A theory of action perspective.* Boston: Addison-Wesley.

Armstrong, A., & Hagel, J. (1996). The real value of on-line communities. *Harvard Business Review, 74*(3), 134–141. https://doi.org/10.1016/S0099-1333(97)90159-2.

Attaran, M. (2004). Exploring the relationship between information technology and business process reengineering. *Information and Management, 41*(5), 585–596. https://doi.org/10.1016/S0378-7206(03)00098-3.

Autor, D. H. (2015). Why are there still so many jobs? The history and future of workplace automation. *Journal of Economic Perspectives, 29*(3), 3–30. https://doi.org/10.1257/jep.29.3.3.

Bagozzi, R. P., & Dholakia, U. M. (2006). Antecedents and purchase consequences of customer participation in small group brand communities. *International Journal of Research in Marketing, 23*(1), 45–61. https://doi.org/10.1016/j.ijresmar.2006.01.005.

Bakhshi, H., Downing, J. M., Osborne, M. A., & Schneider, P. (2017). *The future of skills. Employment in 2030.* London: Pearson and Nesta.

Baldwin, C., Hienerth, C., & von Hippel, E. (2006). How user innovations become commercial products: A theoretical investigation and case study. *Research Policy, 35*(9), 1291–1313. https://doi.org/10.1016/j.respol.2006.04.012.

Baskerville, R., & Dulipovici, A. (2006). The theoretical foundations of knowledge management. *Knowledge Management Research and Practice, 4,* 2. https://doi.org/10.1057/palgrave.kmrp.8500090.

Bathelt, H., Malmberg, A., & Maskell, P. (2004). Clusters and knowledge: Local buzz, global pipeline and the process of knowledge creation. *Progress in Human Geography, 28*(1), 32–56.

Bettiol, M., Capestro, M., Di Maria, E., & Furlan, A. (2019). *Impacts of industry 4.0 investments on firm performance. Evidence From Italy* (No. 233). Retrieved from https://www.economia.unipd.it/sites/economia.unipd.it/files/20190233.pdf

Boden, M. A. (1998). Creativity and artificial intelligence. *Artificial Intelligence, 103*, 347–356.
Boden, M. A. (2016). *AI its nature and future*. Oxford: Oxford University Press.
Bogers, M., Hadar, R., & Bilberg, A. (2016). Additive manufacturing for consumer-centric business models: Implications for supply chains in consumer goods manufacturing. *Technological Forecasting and Social Change, 102*, 225–239. https://doi.org/10.1016/j.techfore.2015.07.024.
Bonilla, S. H., Silva, H. R. O., Terra Da Silva, M., Gonçalves, R. F., & Sacomano, J. B. (2018). Industry 4.0 and sustainability implications: A scenario-based analysis of the impacts and challenges. *Sustainability, 10*(10), 3740. https://doi.org/10.3390/su10103740.
Boschma, R. A. (2005). Proximity and innovation: A critical assessment. *Regional Studies, 39*(1), 61–74. https://doi.org/10.1080/0034340052000320887.
Boudreau, K. J., & Lakhani, K. R. (2009). How to manage outside innovation. *MIT Sloan Management Review, 50*(4), 69–76.
Brown, J. S., & Duguid, P. (2000). Balancing act: How to capture knowledge without killing it. *Harvard Business Review, 78*(3), 73–80.
Brown, J. S., & Duguid, P. (2001). Knowledge and organization: A social-practice perspective. *Organization Science, 12*(2), 198–213. https://doi.org/10.1287/orsc.12.2.198.10116.
Brown, P., Bocken, N., & Balkenende, R. (2019). Why do companies pursue collaborative circular oriented innovation? *Sustainability, 11*(3), 635. https://doi.org/10.3390/su11030635.
Büchi, G., Cugno, M., & Castagnoli, R. (2020). Smart factory performance and Industry 4.0. *Technological Forecasting and Social Change, 150*(June 2019), 119790. https://doi.org/10.1016/j.techfore.2019.119790.
Chesbrough, H. W. (2003). *Open innovation: The new imperative for creating and profiting from technology*. Boston: Harvard Business School Press. https://doi.org/10.1111/j.1467-8691.2008.00502.x.
Coase, R. H. (1937). The nature of the firm. *Economica, 4*(16), 386–405. https://doi.org/10.1111/j.1468-0335.1937.tb00002.x.
Cova, B., & Dalli, D. (2009). Working consumers: The next step in marketing theory? *Marketing Theory, 9*(3), 315–339. https://doi.org/10.1177/1470593109338144.
Cowan, R., David, P. A., & Foray, D. (2000). The explicit economics of knowledge codification and tacitness. *Industrial and Corporate Change, 9*(2), 211–253. https://doi.org/10.1093/icc/9.2.211.
Cyert, M., & March, J. G. (1963). *A behavioral theory of the firm*. New Jersey, NJ: Prentice-Hall.
Dachs, B., Kinkel, S., & Jäger, A. (2019). Bringing it all back home? Backshoring of manufacturing activities and the adoption of Industry 4.0 technologies. *Journal of World Business, 54*(6), 101017. https://doi.org/10.1016/j.jwb.2019.101017.
Daugherty, P., & Wilson, H. J. (2018). *Human + machine: Reimagining work in the age of AI*. Cambridge, MA: Harvard Business School Press.
Davenport, T. (1994). Saving IT's soul: Human-centered information management. *Harvard Business Review, 72*(2), 119–131.
Davenport, T. H. (2007). Information technologies for knowledge management. In K. Ichijo & I. Nonaka (Eds.), *Knowledge creation and management: New challenges for managers* (pp. 97–120). New York: Oxford University Press.
Davenport, T. H. (2018a). From analytics to artificial intelligence. *Journal of Business Analytics, 1*(2), 73–80. https://doi.org/10.1080/2573234x.2018.1543535.
Davenport, T. H. (2018b). *The AI advantage: How to put the artificial intelligence revolution to work*. Boston, MA: MIT. https://doi.org/10.7551/mitpress/11781.001.0001.
Davenport, T., & Prusak, L. (2000). *Working knowledge: Managing what your organization knows*. Cambridge, MA: Harvard Business School Press. https://doi.org/10.1145/348772.348775.
De Marchi, V. (2012). Environmental innovation and R&D cooperation: Empirical evidence from Spanish manufacturing firms. *Research Policy, 41*(3), 614–623. https://doi.org/10.1016/j.respol.2011.10.002.

Despeisse, M., Baumers, M., Brown, P., Charnley, F., Ford, S. J., Garmulewicz, A., et al. (2017). Unlocking value for a circular economy through 3D printing: A research agenda. *Technological Forecasting and Social Change, 115*, 75–84. https://doi.org/10.1016/j.techfore.2016.09.021.

Di Bernardo, B., Grandinetti, R., & Di Maria, E. (2012). Exploring knowledge-intensive business services. In E. Di Maria, R. Grandinetti, & B. Di Bernardo (Eds.), *Exploring knowledge-intensive business services: Knowledge management strategies*. London: Palgrave Macmillan. https://doi.org/10.1057/9781137008428.

Di Maria, E., & Finotto, V. (2008). Communities of consumption and made in Italy. *Industry and Innovation, 15*(2), 179–197. https://doi.org/10.1080/13662710801954583.

Drucker, P. F. (1999). Knowledge-worker productivity: The biggest challenge. *California Management Review, 42*(1), 79–94.

Floridi, L. (2005). Consciousness, agents and the knowledge game. *Minds and Machines, 15*(3–4), 415–444. https://doi.org/10.1007/s11023-005-9005-z.

Franke, N., & Piller, F. (2004). Value creation by toolkits for user innovation and design: The case of the watch market. *Journal of Product Innovation Management, 21*(6), 401–415. https://doi.org/10.1111/j.0737-6782.2004.00094.x.

Fueller, J., & Hippel, E. Von. (2008). *Costless creation of strong brands by user communities: Implications for producer-owned brands* (MIT Sloan School Working Paper, 08), pp. 1–30. https://doi.org/10.2139/ssrn.1275838.

Füller, J., Jawecki, G., & Mühlbacher, H. (2006). Innovation creation by online basketball communities. *Journal of Business Research, 60*(1), 60–71. https://doi.org/10.1016/j.jbusres.2006.09.019.

Gable, G. G., Scott, J. E., & Davenport, T. D. (October, 1998). Cooperative ERP life-cycle knowledge management. In *Proceedings of the Ninth Australiasian Conference on Information System* (pp. 227–240).

Galvagno, M., & Dalli, D. (2014). Theory of value co-creation: A systematic literature review. *Managing Service Quality, 24*(6), 643–683. https://doi.org/10.1108/MSQ-09-2013-0187.

Grant, R. M. (1996). Toward a knowledge-based theory of the firm. *Strategic Management Journal, 17*, 109–122.

Gruner, K. E., & Homburg, C. (2000). Does customer interaction enhance new product success? *Journal of Business Research, 49*(1), 1–14. https://doi.org/10.1016/S0148-2963(99)00013-2.

Gulati, R., & Garino, J. (2000). Get the right mix of bricks & clicks. *Harvard Business Review, 78*(3), 107–114, 214. https://doi.org/10.1057/jors.1978.20.

Hagiu, A., & Wright, J. (January–February, 2020). When data creates competitive advantage. *Harvard Business Review*. Retrieved from https://hbr.org/2020/01/when-data-creates-competitive-advantage

Halbinger, M. A. (July, 2018). The role of makerspaces in supporting consumer innovation and diffusion: An empirical analysis. *Research Policy, 10*, 1–9. https://doi.org/10.1016/j.respol.2018.07.008.

Hammer, M., & Champy, J. (2001). *Reengineering the company – A manifesto for business revolution*: Vol. 19. New York, USA: Harper Business. https://doi.org/10.5465/AMR.1994.9412271824

Hansen, M., Nohria, N., & Tierney, T. (1999). What's your strategy for managing knowledge? *Harvard Business Review, 72*(2), 106–116.

Holmström, J., Holweg, M., Khajavi, S. H., & Partanen, J. (2016). The direct digital manufacturing (r)evolution: Definition of a research agenda. *Operations Management Research, 9*(1–2), 1–10. https://doi.org/10.1007/s12063-016-0106-z.

Iansiti, M., & Lakhani, K. R. (November, 2014). Digital ubiquity: How connections, sensors, and data are revolutionizing business. *Harvard Business Review, 92*(11), 91–99. https://doi.org/10.1017/CBO9781107415324.004.

Inkpen, A. C. (1996). Creating knowledge through collaboration. *California Management Review, 39*, 1. https://doi.org/10.2307/41165879.

Jensen, M. B., Johnson, B., Lorenz, E., & Lundvall, B. Å. (2007). Forms of knowledge and modes of innovation. *Research Policy, 36*(5), 680–693. https://doi.org/10.1016/j.respol.2007.01.006.

Kalva, R. S. (2015). 3D printing-the future of manufacturing (the next industrial revolution). *International Journal of Innovations in Engineering and Technology, 5*(5), 184–190.

Kamble, S., Gunasekaran, A., & Dhone, N. C. (2020). Industry 4.0 and lean manufacturing practices for sustainable organisational performance in Indian manufacturing companies. *International Journal of Production Research, 58*(5), 1319–1337. https://doi.org/10.1080/00207543.2019.1630772.

Kaplan, A., & Haenlein, M. (2019). Digital transformation and disruption: On big data, blockchain, artificial intelligence, and other things. *Business Horizons, 62*(6), 679–681. https://doi.org/10.1016/j.bushor.2019.07.001.

Kelly, K. (1998). *New rules for the new economy.* New York: Viking.

Kogut, B., & Zander, U. (1996). What firms do? Coordination, identity, and learning. *Organization Science, 7*(5), 502–518. https://doi.org/10.1287/orsc.7.5.502.

Kohtala, C., & Hyysalo, S. (2015). Anticipated environmental sustainability of personal fabrication. *Journal of Cleaner Production, 99*, 333–344. https://doi.org/10.1016/j.jclepro.2015.02.093.

Kozinets, R. V. (1999). E-tribalized marketing?: The strategic implications of virtual communities of consumption. *European Management Journal, 17*(3), 252–264. https://doi.org/10.1016/S0263-2373(99)00004-3.

Kozinets, R. V., Hemetsberger, A., & Schau, H. J. (2008). The wisdom of consumer crowds: Collective innovation in the age of networked marketing. *Journal of Macromarketing, 28*(4), 339–354. https://doi.org/10.1177/0276146708325382.

Lacy, P., & Rutqvist, J. (2016). *Waste to wealth: The circular economy advantage.* https://doi.org/10.1057/9781137530707.

Laplume, A., Anzalone, G. C., & Pearce, J. M. (2016). Open-source, self-replicating 3-D printer factory for small-business manufacturing. *International Journal of Advanced Manufacturing Technology, 85*(1–4), 633–642. https://doi.org/10.1007/s00170-015-7970-9.

Laursen, K., & Salter, A. (2006). Open for innovation: The role of openness in explaining innovation performance among U. K. manufacturing firms. *Strategic Management Journal, 27*(2), 131–150. https://doi.org/10.1002/smj.507.

Liao, S. H. (2003). Knowledge management technologies and applications – Literature review from 1995 to 2002. *Expert Systems with Applications, 25*(2), 155–164. https://doi.org/10.1016/S0957-4174(03)00043-5.

Liebowitz, J. (2001). Knowledge management and its link to artificial intelligence. *Expert Systems with Applications, 20*, 1–6.

Lipparini, A., Lorenzoni, G., & Ferriani, S. (2014). From core to periphery and back: A study on the deliberate shaping of knowledge flows in interfirm dyads and networks. *Strategic Management Journal, 35*(4), 578–595. https://doi.org/10.1002/smj.2110.

March, J. G. (1991). Exploration and exploitation in organizational learning. *Organization Science, 2*(1), 71–87.

Marchi, G., Giachetti, C., & De Gennaro, P. (2011). Extending lead-user theory to online brand communities: The case of the community Ducati. *Technovation, 31*(8), 350–361. https://doi.org/10.1016/j.technovation.2011.04.005.

McAfee, A. (2004). Do you have too much IT? *MIT Sloan Management Review, 45*(3), 18–22.

McKinsey Global Institute. (June, 2015). *The internet of things: Mapping the value beyond the hype* (p. 144). Chicago, IL: McKinsey Global Institute.

Micelli, S. (2017). Le tre rivoluzioni del management digitale. *Sinergie, 35*(103), 13–22. https://doi.org/10.7433/s103.2017.01.

Mittal, S., Khan, M. A., Romero, D., & Wuest, T. (2018). A critical review of smart manufacturing & Industry 4.0 maturity models: Implications for small and medium-sized enterprises (SMEs). *Journal of Manufacturing Systems, 49*(June), 194–214. https://doi.org/10.1016/j.jmsy.2018.10.005.

Nambisan, S., & Nambisan, P. (2008). How to profit from a better "virtual customer environment". *MIT Sloan Management Review, 49*(3), 53–61. https://doi.org/10.2752/175470811X13071166525216.

Nevo, S., & Wade, M. R. (2010). The formation and value of IT-enabled resources: Antecedents and consequences of synergistic relationships. *MIS Quarterly, 34*(1), 163–183.

Nonaka, I. (2000). A firm as a knowledge-creating entity: A new perspective on the theory of the firm. *Industrial and Corporate Change, 9*(1), 1–20. https://doi.org/10.1093/icc/9.1.1.

Nonaka, I., & Takeuchi, H. (1995). *The knowledge-creating company: How Japanese companies create the dynamics of innovation.* Oxford: Oxford University Press.

Parker, G. G., Van Alstyne, M. W., & Choudary, S. P. (2016). *Platform revolution: How networked markets are transforming the economy – And how to make them work for you.* New York: W. W. Norton & Company.

Paschen, J., Kietzmann, J., & Kietzmann, T. C. (2019). Artificial intelligence (AI) and its implications for market knowledge in B2B marketing. *Journal of Business and Industrial Marketing.* 34, no 7, p. 1410-1419 10.1108/JBIM-10-2018-0295

Pauleen, D. J. (2017). Dave Snowden on KM and big data/analytics: Interview with David J. Pauleen. *Journal of Knowledge Management, 21*(1), 12–17. https://doi.org/10.1108/JKM-08-2016-0330.

Pauleen, D. J., & Wang, W. Y. C. (2017). Does big data mean big knowledge? KM perspectives on big data and analytics. *Journal of Knowledge Management, 21*(1), 1–6. https://doi.org/10.1108/JKM-08-2016-0339.

Pfeffer, J., & Sutton, R. I. (2000). *The knowing-doing gap: How smart companies turn knowledge into action.* Cambridge, MA: Harvard Business School Press.

Pfeiffer, S. (2016). Robots, Industry 4.0 and humans, or why assembly work is more than routine work. *Societies, 6*(2), 16. https://doi.org/10.3390/soc6020016.

Porter, M. E. (2001). Strategy and the internet. *Harvard Business Review, 79*(3), 62–78.

Porter, M. E., & Heppelmann, J. E. (2014). How smart, connected products are transforming competition. *Harvard Business Review, 92*(11), 64–88.

Ransbotham, S., Khodabandeh, S., Fehling, R., LaFountain, B., & Kiron, D. (2019). Winning with AI. *MIT Sloan Management Review*, (61180). Retrieved from https://sloanreview.mit.edu/projects/winning-with-ai/

Rauch, E., Dallasega, P., & Matt, D. T. (2016). The way from lean product development (LPD) to smart product development (SPD). *Procedia CIRP, 50,* 26–31. https://doi.org/10.1016/j.procir.2016.05.081.

Rehnberg, M., & Ponte, S. (2018). From smiling to smirking? 3D printing, upgrading and the restructuring of global value chains. *Global Networks, 18*(1), 57–80. https://doi.org/10.1111/glob.12166.

Reinhard, G., Jesper, V., & Stefan, S. (2016). *Industry 4.0: Building the digital enterprise.* Global Industry 4.0 Survey, 1–39. https://doi.org/10.1080/01969722.2015.1007734.

Rheingold, H. (1993). *The virtual community: Finding connection in a computerized world.* Boston, MA: Addison-Wesley Longman Publishing.

Rocca, G. L. (2012). Knowledge based engineering: Between AI and CAD. Review of a language based technology to support engineering design. *Advanced Engineering Informatics, 26*(2), 159–179. https://doi.org/10.1016/j.aei.2012.02.002.

Sanders, A., Elangeswaran, C., & Wulfsberg, J. (2016). Industry 4.0 implies lean manufacturing: Research activities in Industry 4.0 function as enablers for lean manufacturing. *Journal of Industrial Engineering and Management, 9*(3), 811. https://doi.org/10.3926/jiem.1940.

Sawhney, M., Verona, G., & Prandelli, E. (2005). Collaborating to create: The internet as a platform for customer engagement in product innovation. *Journal of Interactive Marketing, 19*(4), 4–34. https://doi.org/10.1002/dir.20046.

Schau, H. J., Muñiz, A. M., & Arnould, E. J. (2009). How brand community practices create value. *Journal of Marketing, 73*(5), 30–51. https://doi.org/10.1509/jmkg.73.5.30.

Schneider, P. (2018). Managerial challenges of Industry 4.0: An empirically backed research agenda for a nascent field. *Review of Managerial Science, 12,* 3. https://doi.org/10.1007/s11846-018-0283-2.

Spender, J.-C. (1996). Making knowledge the basis of a dynamic theory of the firm. *Strategic Management Journal, 17*(S2), 45–62. https://doi.org/10.1002/smj.4250171106.

Strange, R., & Zucchella, A. (2017). Industry 4.0, global value chains and international business. *Multinational Business Review, 25*(3), 174–184. https://doi.org/10.1108/MBR-05-2017-0028.

Szulanski, G. (2000). The process of knowledge transfer: A diachronic analysis of stickiness. *Organizational Behavior and Human Decision Processes, 82*(1), 9–27. https://doi.org/10.1006/obhd.2000.2884.

Tian, X. (2017). Big data and knowledge management: A case of déjà vu or back to the future? *Journal of Knowledge Management, 21*(1), 113–131. https://doi.org/10.1108/JKM-07-2015-0277.

Tseng, M., Tan, R. R., Chiu, A. S. F., Chien, C., & Chi, T. (2018). Resources, Conservation & Recycling Circular economy meets Industry 4.0: Can big data drive industrial symbiosis? *Resources, Conservation & Recycling, 131*(December 2017), 146–147. https://doi.org/10.1016/j.resconrec.2017.12.028.

Unruh, G. (2019). The killer app for 3D printing? The circular economy. *MIT Sloan Management Review, 60*(2), 12–13.

Ustundag, A., & Cevikcan, E. (2018). *Industry 4.0: Managing the digital transformation.* 10.1007/978-3-319-57870-5

Von Hippel, E. (1978). A customer-active paradigm for industrial product idea generation. *Research Policy, 7*(3), 240–266. https://doi.org/10.1016/0048-7333(78)90019-7.

von Hippel, E. (1986). Lead users: A source of novel product concepts. *Management Science, 32*(7), 791–805. https://doi.org/10.1287/mnsc.32.7.791.

von Hippel, E. (1994). "Sticky information" and the locus of problem solving: Implications for innovation. *Management Science, 40*(4), 429–439. https://doi.org/10.1287/mnsc.40.4.429.

Von Hippel, E. (2001). Perspective: User toolkits for innovation. *Journal of Product Innovation Management, 18*(4), 247–257. https://doi.org/10.1016/S0737-6782(01)00090-X.

Webster, K., & MacArthur, E. (2017). *The circular economy: A wealth of flows* (2nd ed.). COwes: Ellen MacArthur Foundation Publishing.

Wiig, K. M. (1999). What future knowledge management users may expect. *Journal of Knowledge Management, 3*(2), 155–166. https://doi.org/10.1108/13673279910275611.

Womack, J. P., & Jones, D. T. (1997). Lean thinking—Banish waste and create wealth in your corporation. *Journal of the Operational Research Society, 48*(11), 1148–1148. https://doi.org/10.1057/palgrave.jors.2600967.

World Economic Forum. (January, 2016). The future of jobs employment, skills and workforce strategy for the fourth industrial revolution. *Growth Strategies*, 2–3. 10.1177/1946756712473437

Wu, X., & Gereffi, G. (2018). Amazon and Alibaba: Internet governance, business models, and internationalization strategies. In R. van Tulder, A. Verbeke, & L. Piscitello (Eds.), *International business in the information and digital age: Vol. 13 Progress in international business research* (pp. 327–356). Bingley: Emerald Group. https://doi.org/10.1108/S1745-886220180000013014.

Xu, L., Wang, C., Luo, X., & Shi, Z. (2006). Integrating knowledge management and ERP in enterprise information systems. *Systems Research and Behavioral Science, 23*(2), 147–156. https://doi.org/10.1002/sres.750.

Yao, X., Zhou, J., Zhang, J., & Boer, C. R. (2017). From intelligent manufacturing to smart manufacturing for Industry 4.0 driven by next generation artificial intelligence and further on. In *2017 5th international conference on enterprise systems (ES)* (pp. 311–318). IEEE. https://doi.org/10.1109/ES.2017.58.

Zawadzki, P., & Żywicki, K. (2016). Smart product design and production control for effective mass customization in the Industry 4.0 concept. *Management and Production Engineering Review, 7*(3), 105–112. https://doi.org/10.1515/mper-2016-0030.

Zhuang, Y.-t., Wu, F., Chen, C., & Pan, Y.-h. (2017). Challenges and opportunities: From big data to knowledge in AI 2.0. *Frontiers of Information Technology and Electronic Engineering, 18*(1), 3–14. https://doi.org/10.1631/FITEE.1601883.

Industry 4.0 and Knowledge Management: A Review of Empirical Studies

Mauro Capestro and Steffen Kinkel

Abstract The recent Industry 4.0 paradigm is revolutionizing the manufacturing processes, the way companies create value and interact with suppliers and customers. The new technologies allow manufacturing companies to gather huge amounts of data that they can use to tailor production, develop customized products and services, as well as improve operation activities in terms of efficiency, productivity, and flexibility. In this new technological scenario, new digital skills and competences (i.e., data management) become strategically important as they could assure new knowledge manufacturing companies to achieve superior competitive advantage. Such new knowledge depends not only on the use of Industry 4.0 technologies but also on the interactions with suppliers and customers as well as on the upgrading of employees' competences. With the aim of deepening the understanding of these dynamics, the chapter reviews the empirical studies related to the adoption of Industry 4.0, by highlighting the role of knowledge management.

1 Introduction

The current manufacturing landscape requires key factors such as efficiency, flexibility, faster responsiveness to market changes and customer demand, as well as a higher focus on product quality and customization that are essential for the survival of manufacturing companies (Almada-Lobo, 2015). Moreover, to compete successfully manufacturing firms need higher level of digitization and automation that means an extensive connectivity between manufacturing processes and other business areas—that is high integration of operation systems with the overall organization structure (Berman, 2012; Rashid & Tjahjono, 2016). In addition to internal

---

M. Capestro (✉)
Department of Economics and Management 'Marco Fanno', University of Padova, Padova, Italy
e-mail: mauro.capestro@unipd.it

S. Kinkel
University of Applied Sciences, Karlsruhe, Germany

integration, manufacturing companies also need higher integration with the external environment and specifically with suppliers and customers (Hakanen & Jaakkola, 2012).

In this sense, the new technological revolution known as Industry 4.0 plays a key role, since it enables the integration between manufacturing operations systems and new Information and Communications Technologies (ICT)—such as cloud computing, big data analytics, Artificial Intelligence (AI), and Internet of Things (IoT)—creating the so-called cyber-physical systems (CPS) (Dalenogare, Benitez, Ayala, & Frank, 2018). Essentially, Industry 4.0 highlights the usage of new technologies with the integration among objects, humans, and machines across organizational boundaries to create not only a cyber-physical manufacturing environment but also a new type of networked value chain (Kagermann, 2015). Through the Industry 4.0 technologies, manufacturing companies may be able to implement three types of integrations: (1) horizontal, (2) vertical, and (3) end-to-end integration, which allow them to improve both the operation (efficiency, productivity, quality, etc.) and the marketing (new product development, customization, time-to-market response, etc.) activities (Schwab, 2017).

Although most of the Industry 4.0 technologies are not completely new, the full potential of these technologies has not been exploited in the current manufacturing system (Fatorachian & Kazemi, 2018) due to the higher integration of different information systems previously limited to single business areas (Panetto & Molina, 2008) and the presence of many Industry 4.0 technologies. Industry 4.0 technologies also offer the opportunity to gather and manage a huge amount of data used to improve the firm's offering, both in terms of efficiency and response to customer needs (Büchi, Cugno, & Castagnoli, 2020). In this perspective, knowledge becomes an essential variable that manufacturing companies should consider and manage in an effective way for the successful implementation of Industry 4.0 (Feng, Bernstein, Hedberg, & Barnard Feeney, 2017). In particular, Industry 4.0 technologies enable companies to create new knowledge coming from the use of new technologies in production processes, but also from external environment and precisely through the acquisition and elaboration of data gathered from suppliers and through higher involvement of customers that become active partner (co-creation) in the design process (Lu, 2017). Moreover, knowledge also becomes essential in terms of skills and competences that employees and managers should have to get the highest potential from Industry 4.0, as the new knowledge created should be used in a timely manner (Ardito, Messeni Petruzzelli, Panniello, & Garavelli, 2019).

Therefore, manufacturing companies should take into relevant consideration the strategic relationship between new technologies and knowledge management for the successful implementation of Industry 4.0 technologies. Indeed, this aspect represents one of the main challenges manufacturing companies must overcome to implement successfully Industry 4.0 (Whysall, Owtram, & Brittain, 2019). In this respect, the purpose of this chapter is to highlight the role of knowledge in the Industry 4.0 paradigm by means of literature review of the empirical studies focused on the implementation of Industry 4.0 technologies.

2 Theoretical Background

2.1 The Industry 4.0 Paradigm

Industry 4.0, also known as "fourth industrial revolution," is a current hot topic in both professional and academic fields (Liao, Deschamps, Loures, & Ramos, 2017). Industry 4.0 is a disruptive phenomenon that is changing the rules of competition, allowing companies taking advances in different business processes. Specifically, it is affecting the firms' overall strategy, changing organizational mindset, business models, value creation process, supply chain activities, production processes, products, skills, and stakeholder relationships (Reinhard, Jesper, & Stefan, 2016).

The term *Industry 4.0* comes from Germany as *Industrie 4.0* being a part of the "High-Tech Strategy 2020 for Germany" (Kagermann, Lukas, & Wahlster, 2011). Germany introduced such a term when the government decided to promote and support the technological integration of manufacturing plants with products and business processes to strengthen Germany's position as a leading manufacturing power worldwide (Kagermann, Helbig, & Wahlster, 2013; Lasi, Kemper, Fettke, Feld, & Hoffmann, 2014). Before other countries, the German Government was confident that the future of manufacturing industries will be characterized by a strong product personalization under the conditions of highly flexible production and with the extensive integration of business partners and customers in value creation and business models (Thoben, Wiesner, & Wuest, 2017). Later, besides Germany, other national and international institutions started to give importance to the new phenomenon supporting the implementation of the new technologies. The European Union promoted a public–private partnership under the name *Factories of the Future* to sustain the competitiveness of production. The Italian Government launched in 2016 the *Industry 4.0 National Plan*. France promoted the implementation of Industry 4.0 through the *Aliance Industrie du Futur* initiative and the *Future of Manufacturing* in the United Kingdom (Büchi et al., 2020; da Silva, Kovaleski, & Negri Pagani, 2019). Moreover, in the USA similar efforts are underway through the *Industrial Internet Consortium*. In China, the *Internet Plus initiative* and *Made in China 2025* represent technological initiatives similar to Industry 4.0 (Müller & Voigt, 2018). In addition to China, the other Asian countries that proposed similar initiatives to facilitate the development of their own manufacturing industries are Japan with the *Connected Industries* and Korea with *Smart Factory* (da Silva et al., 2019; Mittal et al., 2019).

Notwithstanding the specific technological representation of the fourth industrial revolution, also as a consequence of the several worldwide initiatives, a common definition of Industry 4.0 has not been accepted. Indeed, both scholars and practitioners in addition to the governments used several synonyms to depict it. In particular, *Industry 4.0* (and/or *Industrie 4.0*) is predominantly used in Europe. *Smart manufacturing* is the term predominantly used in the USA and *smart factory* in Asia (Kagermann, Gausemeier, Schuh, & Wahlster, 2016; Mittal, Khan, Romero, & Wuest, 2018). Moreover, other terms most used in the business management field are *Cyber-physical system* (CPS), *(Industrial) Internet of Things* (IoT/IIoT), *digital manufacturing/factory*, and i*ntelligent manufacturing systems* (IMS) (Agostini & Filippini, 2019; Büchi et al., 2020; Thoben et al., 2017). Table 1 reports a description of Industry 4.0 definitions.

Despite some differences in their definitions, the terms can be considered interchangeably. Most of them focus principally on manufacturing and operation activities, but all of them highlight the digitization of business activities (Müller, Buliga, & Voigt, 2018). Other common elements regard the automation systems, the connections between the physical and digital worlds, the extensive use of Internet, and the changes in the relationships with stakeholders and in business governance (Büchi et al., 2020).

Industry 4.0 is a broader concept that considers all the different aspects emerged from the abovementioned definitions (see Mohamed, 2018). It encompasses the adoption of industrial automation systems that support companies in managing the production and value creation processes, the supply chain activities, and all their related processes (Reischauer, 2018; Yin, Stecke, & Li, 2017). One of the most accepted definition of Industry 4.0 is provided by McKinsey Company that considers it as the digitization of the manufacturing sector, with embedded sensors in virtually all product components and manufacturing equipment, ubiquitous cyber-physical systems, and analysis of all relevant data (Wee, Kelly, Cattel, & Breunig, 2015). More generically, Industry 4.0 can be defined as a new business approach for controlling production processes by providing real-time data analysis and by enabling the unitary customization of products (Kohler & Weisz, 2016).

Thus, the concept of Industry 4.0 is mainly embedded in smart manufacturing and CPS, whose technological infrastructure bases on the concept of IoT that allows connection of machines, products, systems, and people (Kagermann et al., 2013; Lasi et al., 2014; Schmidt et al., 2015). Together with cloud computing, CPS and IoT are considered as the central pillars of Industry 4.0 (Müller, 2019a). Finally, the integration of CPS with innovative production, logistics, and services practices led to the Industry 4.0 factory (Lee, Bagheri, & Kao, 2015). Indeed, Industry 4.0 is often identified with a set of technologies that bring together the CPS domain (IoT, big data and analytics, cybersecurity) with other production technologies such as advanced manufacturing systems, additive manufacturing, 3D printing, horizontal/vertical integration, and simulation systems (Agostini & Filippini, 2019; Tortorella & Fettermann, 2018).

2.2 *The Industry 4.0 Enabling Technologies*

The implementation of Industry 4.0 paradigm in the manufacturing sectors covers a wide range of applications from product development and design to operation and logistic activities. For this reason, Industry 4.0 comprises a very large set (estimated more than 1200) of enabling technologies (Chiarello, Trivelli, Bonaccorsi, & Fantoni, 2018), even if scholars and practitioners focused only on some enabling

Table 1 Industry 4.0 definitions

Term	Definition	Author(s)
Smart manufacturing	Smart Manufacturing is the integration of information technology (IT) and data with different manufacturing technologies, processes, and resources to enable intelligent, efficient, and responsive operations. It relies on the digitalization and interconnection of entire value chains, products, processes, and business models	Kagermann (2015), Thoben et al. (2017), Kusiak (2018), Lu and Weng (2018), Mittal et al. (2019)
Smart factory	A smart factory is a factory that is context-aware and assists people and machines in execution of their tasks by systems working in background. It differs from smart manufacturing as it focuses on a single plant rather than on a broader supply network concept	Hozdić (2015), Jung, Choi, Kulvatunyou, Cho, and Morris (2017), Mittal et al. (2018), Prause (2019)
Cyber-physical (production) system (CPS)	CPS refers to the use of technologies for the management of interconnected systems between its physical assets and computational capabilities. It focuses on the data gathered, stored, and shared to operate autonomously	Schlechtendahl, Keinert, Kretschmer, Lechler, and Verl (2015), Agostini and Filippini (2019), Müller, Buliga, et al. (2018), Müller (2019a)
(Industrial) Internet of Things (IoT/IIoT)	IoT and IIoT rely on the digital connection between physical entities and digital components for a completely intelligent, inter-connected, and autonomous factory. It results not only in a production change but in an extensive organizational change	Arnold, Kiel, and Voigt (2016), Kiel, Arnold, and Voigt (2017), Arnold and Voigt (2019)
Digital manufacturing/ factory	Digital manufacturing/factory concept relies on the use of computer-assisted applications, analytics, simulation, three-dimensional (3D) visualization, and various collaboration tools, integrated in a common communication infrastructure Data management systems and simulation technologies are concurrently used to optimize the manufacturing processes and match the customer demands	Chen et al. (2015), Byrne et al. (2016), Cavalcante, Frazzon, Forcellini, and Ivanov (2018)
Intelligent manufacturing systems (IMS)	The notion of IMS is quite similar to smart manufacturing, but it stresses the role of control on production activities. It relies on the ability to self-regulate and/or self-control to manufacture the products within the design specifications. Moreover, it relies on a new way of interacting between humans and production machines with the aim of creating a whole smart factory	Zhong, Xu, Chen, and Huang (2017), Zhong, Xu, Klotz, and Newman (2017) and Stadnicka, Litwin, and Antonelli (2019)

technologies considered the pillars of manufacturing technological digitalization (Agostini & Filippini, 2019; Moeuf et al., 2020).

Considering Industry 4.0 as a new manufacturing approach that relies on technologies able to gather and analyze data in real time, to control and customize the production processes, we have limited the scope of our review to empirical studies concerning the adoption of the following enabling technologies (Agostini & Filippini, 2019; Büchi et al., 2020; da Silva et al., 2019; Mittal et al., 2018; Moeuf, Pellerin, Lamouri, Tamayo-Giraldo, & Barbaray, 2018; Moeuf et al., 2020):

- *Industrial Internet of Things*: new digital technologies, devices, and sensors that facilitate communication between people, products, and machines. Internet is the center of connectivity for all the intelligent devices in which real-time communication provides valuable feedbacks and information that improve production process and delivery and facilitate the decentralization of decision-making.
- *Cloud computing*: technologies that facilitate the storage and processing of huge amounts of data with high performance in terms of speed, flexibility, and efficiency. Cloud computing allows in real time information sharing across multiple systems and networks ensuring data for the production system, including monitoring and control functions, and improving the quality of operations.
- *Big data and analytics*: technologies, tools, and techniques useful to capture, archive, analyze, and disseminate large quantities of data derived from products, processes, machines, and people interconnected in a company, as well as the environment around it. Big data and analytics may assure benefits in terms of higher product customization, higher flexibility due to the possibility of demand estimations, a better product quality and less production waste, and the optimization of supply chain.
- *Virtual and Augmented reality*: a series of technologies and devices used to simulate an environment containing real and virtual objects with the aim of improving production processes by enhancing design and prototyping and product development, reducing setup costs and process time, receiving information in real time, and providing virtual training. In this way, the human performances increase through the ability to reproduce and reuse digital information and knowledge to support the operation activities.
- *Additive manufacturing*: the most important example is the 3D printing technology that is grounded in the additive production creating layers by layers the shape of object deriving from a 3D design file. 3D printing can use several different materials, such as plastics, ceramics, metals, and resins and others, eliminating the assembling of the final product. This kind of technology enables companies to improve the design, prototyping and production of complex products, as well as product customization. Additive manufacturing is beneficial for the supply chains where the production of spare parts is a key part of the business due to high-level after-sales services.
- *Artificial intelligence*: automated solutions developed by the massive amounts of data gathered and that are able to act alone without the intervention of humans to solve problems that otherwise require human intervention. It is "a system's ability

to interpret external data correctly, to learn from such data, and to use those learnings to achieve specific goals and tasks through flexible adaptation" (Haenlein & Kaplan, 2019, p. 5). The application of artificial intelligence to production processes can support both productivity and quality taking decisions and carrying out actions based on the evaluation of the current environment.

- *Advanced manufacturing solutions*: interconnected and modular systems by means of robots and embedded sensor technologies that are becoming increasingly flexible, communicative, and cooperative guaranteeing the complete automation of industrial plants. These technologies include different automatic machinery used in the production processes, such as the material-moving systems, autonomous and advanced robotics, the collaborative robots (cobots), and automated guided vehicles or unmanned aerial vehicles. These technologies do not eliminate the workforce but improve the collaboration with them, with new tasks focused mainly on planning and control phases, eliminating the structural and technological constraints of automatic and fixed systems.
- *Horizontal and vertical integration systems*: The integration offered by Industry 4.0 is characterized by two dimensions: internal versus external. The first (horizontal integration) concerns the integration and exchange of information among the different areas in the company. The second (vertical integration) concerns the company's relationships with its suppliers and customers. Integration systems allow us to reduce setup and production costs and, at the same time, to improve product quality due to the better connections in the incoming and outgoing supply chain activities, with positive effects also on productivity.
- *Simulation*: integration of different computer tools to reproduce the physical world in the virtual models allowing all company operators (managers, designers, production workers) to test and optimize the production settings and simulate the performance of production system. Modeling enables the analysis of materials, production line and product performance, as well as the multisite coordination with the aim of improving and optimizing the overall operations and reducing setup costs, errors, and machine downtimes.

Indeed, the two key factors for Industry 4.0's successful implementation are *integration* and *interoperability*. Integration affects networking with stakeholders, both horizontally and vertically. Interoperability helps the production processes, within and beyond the boundaries of the organization, interconnecting systems and exchanging knowledge and skills (Lu, 2017). This means that companies to implement successfully the Industry 4.0 paradigm should be open to organizational changes (Arnold et al., 2016).

The advent of manufacturing digitalization induces, therefore, companies to evolve and change their own structure accepting knowledge-based, dynamic, and collaborative learning governance models, with the implementation of new technologies being extremely knowledge intensive (Ghobakhloo, 2018). The decision-making process of manufacturing companies that implemented Industry 4.0 requires information and knowledge from the data that technologies allow to collect (Feng et al., 2017).

2.3 The Role of Knowledge in the Fourth Industrial Revolution

In the fourth industrial revolution, the main goal of the implementation of new technologies is related to the effective and efficient customer-oriented adaptation of products (and thus production) and services in order to increase the value added for companies, raising their competitive position, while for customers improving satisfaction and loyalty (Roblek, Meško, & Krapež, 2016). In order to achieve this goal, manufacturing companies need to develop and manage new knowledge that is crucial for the organization's decision-making process and the achieving of the related business goals (Abubakar, Elrehail, Alatailat, & Elçi, 2019).

Industry 4.0 reflects a combination of digital and manufacturing technologies which enable vertical integration of the company' systems, horizontal integration in collaborative networks, and end-to-end solutions across the value chain that guarantee automated, flexible, and self-configurable intelligent processes to enable the creation of new revenue sources (Schneider, 2018). Specifically the new technological transformation embraces technological advances that concern the production process (i.e., advanced manufacturing systems, autonomous robots, additive manufacturing), the use of smart products (e.g., IoT and IIoT), and/or data tools and analytics (big data, AI, cloud, etc.) (Porter & Heppelmann, 2014, 2015). Within the manufacturing process, the adoption of autonomous and/or collaborative robotics (Adamson, Wang, & Moore, 2017) or 3D printing is opening up new opportunities to create new knowledge concerning products and processes (Anderson, 2012; Bogers, Hadar, & Bilberg, 2016). At the same time, smart products and "data-driven technologies" enable the successful acquisition of useful data from several sources within the organizational boundaries as well as from customers and suppliers (Klingenberg, Viana Borges, & Valle Antunes Jr., 2019). Therefore, Industry 4.0 stresses the huge potentialities of data that can be used in real time, enriching contextual knowledge or generating new one in the way products can be produced and used, as well as in the practices concerning value generation (from product to service), allowing firms to take actions and make decisions based on such knowledge (Tao, Qi, Liu, & Kusiak, 2018). Moreover, it is essential to consider a holistic view of manufacturing processes, integrating data from different sources to achieve the business benefits of new technologies (Schneider, 2018).

Finally, the huge amount of data produced by the new technologies needs humans for the creation and management of new useful knowledge (Ramzi, Ahmad, & Zakaria, 2018). In this sense, Industry 4.0 opens another issue for the manufacturing companies that choose to implement and use the new technologies successfully, namely the acquisition of new skills and competences (Schwab, 2017). In this case, new knowledge to manage (acquisition, conversion, application) data and processes needs to be acquired by employees and managers (Dragicevic, Ullrich, Tsui, & Gronau, 2019). People have to gain knowledge that will enable the development of *digital thinking* so that they may manage the process in a new way (Schniederjans, Curado, & Khalajhedayati, 2019). The networking of production systems in the smart manufacturing context, for example, relies on *interconnectivity* and *interoperability* across the different intelligent components requiring knowledge of intelligent vertical and horizontal networking. Manufacturing companies with digital skills within their human resources would be more ready to start digital transformation (Gilchrist, 2016). Moreover, the integration between technologies, knowledge, and human competences and capabilities should allow companies to be more ready for digital transformation and improve their competitiveness achieving their strategic goals (Dalenogare et al., 2018).

3 Research Goals and Methodology

Currently, the attention of researchers and practitioners on the implementation of Industry 4.0 focuses on the impacts of new technologies on business processes and performance as well as on the relationships with customers and suppliers to achieve a superior competitive advantage (Wagire, Rathore, & Rakesh, 2019). In this process, the new knowledge companies created through both the use of Industry 4.0 technologies and the new forms of relationships with suppliers and customers (Di Maria, Bettiol, Capestro, & Furlan, 2018; Schniederjans et al., 2019) has a key role (Karia, 2018). Our main goal is to identify the empirical (quantitative and qualitative) studies to better explain the relationship between Industry 4.0 and knowledge management. Specifically, reviewing the recent empirical studies, we analyze more in depth how the manufacturing companies that adopted the Industry 4.0 technologies gather data, create new knowledge for processes (and strategy), products (and related services), and people (workers/customers/suppliers), and use it for the achievement of business results and the improvement of competitive advantage. In doing so, we try to answer the following research questions.

RQ1. How is Industry 4.0 linked to the company's knowledge management process?
RQ2. How do companies create new knowledge within the Industry 4.0 paradigm?
RQ3. How do companies manage and use new knowledge related to Industry 4.0 technologies to sustain their competitive advantage?

To reach the research purposes and answer the abovementioned research questions, we reviewed papers published in scholarly international journals, specifically business and management journals (Durach, Wieland, & Machuca, 2015; Tranfield, Denyer, & Smart, 2003). The literature review focuses on articles dealing exclusively with empirical studies (qualitative and quantitative) on the adoption of Industry 4.0 technologies. In this regard, we selected the papers considering the following searching criteria. Firstly, we considered the papers that have in abstracts, keywords, and title at least one (OR) of the Industry 4.0 definitions before mentioned and specifically: Industry 4.0, Industrie 4.0, smart manufacturing, smart factory, cyber-physical systems, industrial Internet of Things, digital manufacturing, digital factory, and intelligent manufacturing systems. In addition, we also considered (AND) both the presence of keyword "knowledge" and that of at least one of the following keywords (OR), "empirical studies," "survey," "in-depth interview," or "case study." Finally, the articles were searched on Scopus, ScienceDirect, and Web of Science that are the databases most used in the business and engineering management disciplines (Moeuf, Pellerin, Lamouri, Tamayo-Giraldo, & Barbaray, 2018, 2020).

Moreover, taking into consideration the evolving of research and number of publications on Industry 4.0 (Wagire et al., 2019), we focused the analysis on articles (only published articles and not conference proceedings and book chapters) published starting from 2016. As result of this first step of the review process, 251 articles were found. A following analysis of the titles and abstracts of the papers identified enabled a selection of 60 articles excluding those that were incompatible with the research criteria and purposes. A final full-text analysis of the 60 papers selected made it possible to focus the review of empirical studies on 50 articles that enable us to answer the research questions and highlight the role of knowledge in the Industry 4.0 paradigm. Consistently with our theoretical premises, we focused on the studies that investigated the adoption and use of a set or a specific group (i.e., data-driven technologies) of enabling technologies rather than one single technology (i.e., big data or additive manufacturing or IoT, etc. might be explored in future research).

3.1 Results and Discussion

The analysis of the paper selected focused on the implementation of Industry 4.0 technologies and on the relationship with knowledge management. Specifically, we aimed to analyze the empirical studies highlighting the key role of knowledge in the successful implementation of the Industry 4.0 paradigm. In doing so, we focused on the drivers and barriers of the Industry 4.0 adoption as well as on the (economic and organizational) benefits of the new technologies implementation, trying to understand the role of knowledge (sources, value, and management) from the viewpoint of business and production *processes*, *products* and related services, and internal (human resources) and external (suppliers and customers) *people* involved with the new technologies (Cepeda Carrión, Luis Galán González, & Leal, 2004; Santos et al., 2017). Before the presentation of the results of analysis referred to the processes, products, and the people dimensions of Industry 4.0 and knowledge management relationships, Table 2 summarizes the papers analyzed from the viewpoint of technologies, country, research method (qualitative and quantitative), topic, and main findings.

Firstly, Table 2 shows that the number of empirical studies started to grow significantly since 2018. In particular, in the last 2 years the number of surveys has grown significantly. Germany, Italy, and Brazil are the most active countries in terms of empirical studies carried out to explore the phenomenon of Industry 4.0. After a first understanding of the drivers, barriers, and challengers (Feng et al., 2017; Kiel, Müller, Arnold, & Voigt, 2017; Prause & Atari, 2017) related to the new technological revolution and to the implementation patterns, research focused on the

Table 2 Overview of the empirical studies analyzed

Year	Author(s)	Technologies	Country	Method Quali	Method Quanti	Drivers-Barriers	Benefits	Process	Product	People	Main findings
2020	Büchi et al.	Industry 4.0 technologies	Italy		x		x	x	x	x	Integrating industrial automation systems both vertically and horizontally facilitates production processes and knowledge and skills exchange improving production process and outcomes
2020	Kamble et al.	Industry 4.0 technologies	India		x	x	x	x			Industry 4.0 affects sustainable organizational performance (SOP) and lean manufacturing practices (LMP)
2020	Moeuf et al.	Industry 4.0 technologies	International	x		x		x		x	Lack of competency is an organizational challenge Lack of technological and strategical expertise is the main barrier. Training, data management, and exploitation are the main critical success factors
2020	Tortorella et al.	Cloud IoT Big data and analytics	Brazil		x		x	x		x	Organizational capabilities positively mediate the impact of I4.0 for achieving higher operational performance

(continued)

Table 2 (continued)

Year	Author(s)	Technologies	Country	Method		Drivers-Barriers	Benefits	Process	Product	People	Main findings
				Quali	Quanti						
2019	Agostini and Filippini	Industry 4.0 technologies	Italy		x	x		x		x	Organizational and managerial practices and suppliers/customers technological integration affect the implementation of I4.0 technologies
2019	Agostini and Nosella	Industry 4.0 technologies	Italy Germany Austria Poland Hungary		x	x		x		x	Internal and external social capital, management, and prior investment in advanced manufacturing technologies support the adoption and the intensity of I4.0 technologies
2019	Ardito et al.	Industry 4.0 technologies	International	x			x	x			Key role of I4.0 in terms of information acquisition, storage, and knowledge elaboration for supply chain integration
2019	Arnold and Voigt	IIoT	Germany		x	x				x	Company's top management is one of the main drivers of IIoT implementation
2019	Butschan et al.	IIoT	Germany		x	x				x	High developed cognitive and processual competencies promote the company's digital transformation

Year	Author	Topic	Country						Findings
2019	Dachs et al.	Industry 4.0 technologies	Austria Germany Switzerland	x		x	x	x	Positive relationship between backshoring and I4.0 through increased coordination in production process and integration with suppliers with positive impacts on product customization and time to market
2019	Dalmarco et al.	Industry 4.0 technologies	Portugal	x	x	x	x	x	Data management and the integration of new technologies with available ICT are the main challenges. New data allow improvements of production process, product quality and services, and business model innovation through the integration with external partners
2019	Frank et al.	Smart manufacturing IoT Cloud Big data and analytics	Brazil		x	x			I40 implementation is affected by three maturity clusters (vertical integration and traceability; automation; flexibilization) with cloud implemented in the first stage; IoT in the second; and big data and analytics in the third stage

(continued)

Table 2 (continued)

Year	Author(s)	Technologies	Country	Method		Drivers-Barriers	Benefits	Process	Product	People	Main findings
				Quali	Quanti						
2019	Ghouri and Mani	Industry 4.0 technologies	Malaysia		x		x			x	Positive role of cloud technology and real-time information sharing in enhancement of customer experience
2019	Jerman et al.	Smart factory	Slovenia	x			x	x			Technological transformation encourages the creation of new knowledge that affects organization and business model
2019	Kohnová et al.	Industry 4.0 technologies	Slovakia Czech Republic Austria Germany Switzerland		x	x		x	x	x	Industry 4.0 is affected by external and internal factors, such as the education system, external partnerships, corporate culture, and the new knowledge about processes and new products firms are able to develop and share among workers
2019	Mihardjo et al.	Industry 4.0 technologies	Indonesia		x	x		x		x	Digital capabilities to establish good customer experience and to innovate business model
2019	Mittal et al.	Smart manufacturing Industry 4.0 technologies	International	x		x	x	x		x	SMEs may exploit the benefits of automation improving employee skills and data management

Year	Author	Technology	Country							Findings
2019a	Müller	Cyber-physical systems	Germany	x				x		x	Value proposition, customers, and key activities are the main drivers of Industry 4.0 providers. Key partners and customer relationships are the main drivers of Industry 4.0 users
2019b	Müller	IIoT	Germany	x		x		x		x	The main potentials of digital platforms on business and production processes depend on coordination and information
2019	Prause	Big data analytics, cloud, IoT Simulation	Japan		x	x					Market complexity, relative advantage, and top management are the main drivers of adoption
2019	Rajput and Singh	Industry 4.0 technologies	International		x	x	x	x			The I4.0 technologies and the data generated enable firms to achieve eco-efficiency, eco-effectiveness, and eco-design optimizing logistics and manufacturing processes
2019	Rauch et al.	Smart manufacturing	Italy USA Austria Thailand	x		x				x	Drivers of adoption refer to organizational practices, human resource, technological asset, and networking. Barriers refer to organizational culture and strategy, and human and financial resources

(continued)

Table 2 (continued)

Year	Author(s)	Technologies	Country	Method		Drivers-Barriers	Benefits	Process	Product	People	Main findings
				Quali	Quanti						
2019	Schroeder et al.	IoT Product-use data	UK	x		x	x	x	x		Benefit and opportunities relate to the product-use data provided to the network actors. Lack of digital and data processes capability is the main barrier to I4.0 implementation
2019	Seetharaman et al.	IIoT Big data analytics	India		x	x	x	x			Digitalization and advantages of advanced analytics improve business decisions, performance, and connectivity
2019	Sivathanu	IIoT	India		x	x				x	Digital expertise, infrastructure, relative advantage are the main drivers of adoption
2019	Ślusarczyk et al.	Cyber-physical systems Smart factory	Malaysia		x		x	x	x		I4.0 and data generated affect production processes and activities as well as the development of products-related services
2019	Stentoft and Rajkumar	Industry 4.0 technologies	Denmark		x	x	x	x		x	Customer requirements are the most important drivers of adoption. I4.0 technologies affect the company's globalization strategies and in particular with backshoring

Year	Author	Technology	Country							Findings
2019	Szalavetz A	Industry 4.0 technologies	International	x		x	x	x	x	I4.0 improves production and technological capabilities as well as new knowledge development for products development
2019	Whysall et al.	Industry 4.0 technologies	UK	x		x			x	Industry 4.0 requires changes in talent management practices and the nature of work skill-sets (not only technical knowledge but also networked and collaborative interdisciplinary)
2018	Ardolino et al.	IoT Cloud Data analytics	International	x			x	x		Use of cloud, data analytics, and IoT enables firms to improve product-related services and product customization
2018	Basl	Industry 4.0 technologies	Czech republic		x	x	x	x	x	Customer demands, efficiency, and employee's creativity are the most important motivations of adoption
2018	Bienhaus and Haddud	Big data IoT AI	International		x		x	x	x	Industry 4.0 automatizes buying and other operative activities, supports strategic initiatives driven by humans, and improves buyer–supplier relationships

(continued)

Table 2 (continued)

Year	Author(s)	Technologies	Country	Method		Drivers-Barriers	Benefits	Process	Product	People	Main findings
				Quali	Quanti						
2018	Dalenogare et al.	Industry 4.0 technologies	Brazil		x		x	x	x		I4.0 technologies could be (i) product development and (ii) manufacturing with impacts on operation and products. Data technologies are still at a very early stage of adoption
2018	Fettermann et al.	IoT Cloud Augmented reality Data analytics Additive manufacturing	Germany Japan	x			x	x	x		Use of I4.0 technologies allows manufacturing firms to gather data that they use to improve operation performance and products and service development
2018	Lin et al.	Smart manufacturing	China		x	x	x	x			The perceived benefits such as related to the use of advanced production technology affect the adoption as well as several organizational processes
2018	Lugert et al.	Industry 4.0 technologies	International		x		x	x	x		Integration between simulation and lean and the use of data in real time affect production and product development

Year	Authors	Keywords	Country						Findings
2018	Müller and Voigt	IIoT Smart factory	Germany China	x	x	x			x	IIoT enhances the competitiveness of both German and Chinese firms allowing them to respond to customer requests and reduce production time. Both German and Chinese firms consider partners and suppliers skills very important
2018	Müller, Buliga, et al.	Industry 4.0 technologies	Germany	x		x	x	x	x	Industry 4.0 favors cooperation and new value creation sharing production-related data with suppliers and customers. Servitization by means of data leads to innovative business models
2018	Müller, Kiel, et al.	IIoT	Germany		x	x	x		x	Industry 4.0 positively affects operation activities and employee skills are drivers of adoption
2018	Neirotti et al.	Big data IoT Simulation	Italy		x				x	Firm size and the environmental conditions (dynamism; complexity; munificence) have a combined effect on the development of capabilities that affect the successful use of I4.0 technologies

(continued)

Table 2 (continued)

Year	Author(s)	Technologies	Country	Method		Drivers-Barriers	Benefits	Process	Product	People	Main findings
				Quali	Quanti						
2018	Nagy et al.	Industry 4.0 technologies	Hungary	x	x		x	x	x	x	The huge amounts of data produced by devices, systems, and suppliers and customers support production process, new product development, and all value chain activities
2018	Saniuk and Saniuk	Industry 4.0 technologies	Poland		x	x				x	One of the main barriers of adoption is among others the lack of qualified employment and qualified engineering staff
2018	Schneider	Industry 4.0 technologies	Germany	x		x		x		x	Managerial challenges refer to analysis and strategy; planning and implementation; cooperation and networks; business models; human resources
2018	Tortorella and Fettermann	Industry 4.0 technologies	Brazil		x		x	x			Industry 4.0 and lean practices are positively related; thus knowledge developed for one approach favors the implementation of the other one

Year	Author	Topic	Country								Notes
2017	Feng et al.	Smart factory	USA	x		x		x		x	Industry 4.0 requires different levels of knowledge for the different technologies and business process companies have to manage
2017	Kiel, Arnold, et al.	IIoT	Germany	x			x	x	x		Data from technologies allow new products, services, the optimization of production systems, and processes, the integration of customers and suppliers into product and service but also the need for appropriate workforce
2017	Kiel, Müller, et al.	IIoT	Germany	x		x	x	x			IIoT enhances production and resource efficiency, and business model innovation. Technical integration, organizational transformation, and data security are the most important challenges
2017	Krzywdzinski	Industry 4.0 Automation	Germany Eastern Europe		x	x				x	Automation needs of the introduction of new technological skills to manage processes and new product development

(continued)

Table 2 (continued)

Year	Author(s)	Technologies	Country	Method		Drivers-Barriers	Benefits	Process	Product	People	Main findings
				Quali	Quanti						
2017	Prause and Atari	Industry 4.0 technologies	Germany North-Eastern Europe	x		x	x				Industry 4.0 needs of a successful networked approach with agile and responsive supply chain structures and dynamic capabilities
2016	Arnold et al.	IIoT	Germany	x			x	x		x	The production automation requires qualified workforce and key partner networks to the benefit of operation efficiency

Note: Studies are listed *per* year and alphabetic order. Industry 4.0 technologies refer to the main enabling technologies effects of the new technologies on business processes and strategies (Dalenogare et al., 2018; Kamble, Gunasekaran, & Dhone, 2020; Tortorella, Msc Cawley Vergara, Garza-Reyesc, & Sawhneyd, 2020). Scholars focused the attention on the impacts on production processes and operation performance highlighting the effects on efficiency, productivity, and flexibility (Kiel, Arnold, et al., 2017; Müller, Kiel, & Voigt, 2018; Rajput & Singh, 2019), but also on the international production strategies (Dachs, Kinkel, & Jäger, 2019; Stentoft & Rajkumar, 2019). Moreover, scholars analyzed the role of external interactions with suppliers and customers and the integration with them for the development and sustaining of competitive advantage (Agostini & Filippini, 2019; Frank, Dalenogare, & Ayala, 2019; Ghouri & Mani, 2019; Müller, 2019a). Finally, in addition to the relevance in terms of business advances, the empirical studies aimed at understanding the key role of digital skills and capabilities as drivers of adoption and as a necessary condition to get benefits from the use of new technologies (Büchi et al., 2020; Mittal et al., 2019).

In this new technological scenario, data and the new knowledge they allow to create assume a key role for the successful implementation of Industry 4.0 (Dalmarco, Ramalho, Barros, & Soares, 2019; Ferraris, Mazzoleni, & Devalle, 2019; Jerman, Erenda, & Bertoncelj, 2019). Specifically, some specific technologies such as data technologies (IoT, big data, cloud, AI) enable manufacturing companies to create new knowledge that can be used to improve processes, product, and services (Ardolino et al., 2018; Kiel, Arnold, & Voigt, 2017; Seetharaman, Patwa, Saravanan, & Sharma, 2019). New knowledge can also come from external source and, in this case, depends on the interactions with suppliers and customers (Müller, Buliga, et al., 2018; Nagy, Oláh, Erdei, Máté, & Popp, 2018). New knowledge is needed to manage new technologies; indeed, digital skills and capabilities are the most important challenges companies need to face (Schroeder, Ziaee Bigdeli, Galera Zarco, & Baines, 2019; Sivathanu, 2019; Szalavetz, 2019).

For the research purposes, we considered the relationships between Industry 4.0 and knowledge management from the viewpoints of three main domains. Firstly, we considered the organizational *processes*, as the new technologies need to rethink how the organization operates. Secondly, we considered the *products*, as the new technologies enable the development of new customized products and related services. Finally, we focused on *people*, as the new technologies allow the development of new ways of interactions with suppliers and customers but mainly require new skills and competences.

3.2 Industry 4.0 and Knowledge Management for Processes

The concept of Industry 4.0 and the transformation of industrial production were born with the aim of facing global competition and of adapting manufacturing to the ever-changing market requests. These requirements fostered radical advances in current manufacturing processes introducing new technologies such as autonomous robots, additive manufacturing, advanced manufacturing technologies, and simulation, which allow firms to be more competitive in terms of efficiency and productivity (Rojko, 2017). However, Industry 4.0 is a more complex approach based on integration of the business and manufacturing processes and integration of all actors along the value chain (suppliers and customers). In this sense, data and new knowledge are essential to improve all business processes (Agrawal, Schaefer, & Funke, 2018).

The analysis of empirical studies reported in the Table 2 confirms the key value of knowledge coming from data for the improvement of business processes. The main potentials of digital transformation on business and production processes depend on coordination and information (Müller, 2019b). The new knowledge related to the use of new technologies affects several processes (Lin et al., 2018) starting from operation performance as well as the products and service development (Dalenogare et al., 2018; Fettemann et al., 2018). To achieve different benefits, company needs different levels of knowledge related to the different technologies and business process they have to manage (Feng et al., 2017). In this sense, empirical research shows that knowledge developed within the Industry 4.0 paradigm as well as the use of data in real time affects other production approaches such as lean (and vice versa) (Lugert et al., 2018; Tortorella & Fettermann, 2018). New technologies positively affect operation activities, and employee skills are drivers for adoption (Müller, Kiel, et al., 2018), but also the supply chain integration, favored by the opportunity of higher data acquisition, storage, and knowledge elaboration (Ardito et al., 2019). Empirical research also shows the key role of data for the internationalization strategies and, in particular, for the backshoring of production activities performed abroad by means of increased coordination in production process and integration with suppliers, with positive impacts on product customization and time-to-market response (Dachs et al., 2019; Stentoft & Rajkumar, 2019).

More broadly, the technological transformation boosts the creation of new knowledge that improves decision-making process (Seetharaman et al., 2019) and affects both organization processes and business model (Jerman et al., 2019). Data from technologies allow the optimization of production systems and processes, the development of new products, services, and the integration of customers and suppliers into product and services, but also the need for appropriate workforce (Kiel, Arnold, et al., 2017; Kiel, Müller, et al., 2017).

3.3 *Industry 4.0 and Knowledge Management for Product*

Digital transformation aims at automating the manufacturing processes, improving productivity and production efficiency (Holmström, Holweg, Khajavi, & Partanen, 2016) and, at the same time, satisfying the dynamic customer requests through higher product and service personalization (Wang, Hai-Shu Ma, Yang, & Wang, 2017). Knowledge from customers created by means of new interaction ways and the use of smart products (Porter & Heppelmann, 2014) becomes a strategic data source that firms may use to deliver tailored products and services to the market.

Specific industry 4.0 technologies, such as additive manufacturing, allow firms to improve the participation of customers in the design and production processes (Acharya, Singh, Pereirac, & Singh, 2018).

Past empirical studies stress the role of new technologies and of customers on data gathering and on new products and services development (Ślusarczyk et al., 2019). In addition to customers, also interactions and integration with suppliers allow companies developing new products by means of new knowledge created with them (Kiel, Arnold, et al., 2017). The huge amounts of data produced by devices, systems, and suppliers and customers support production process and product quality as well as all the related services (Dalmarco et al. 2019; Nagy et al., 2018). Indeed, Industry 4.0 enables cooperation and new value creation process through the sharing of production-related data with suppliers and customers. Use of cloud, big data and analytics, and IoT enable firms to improve product-related services and product customization (Ardolino et al., 2018). Therefore, products and services are highly influenced by this new industrial paradigm. Products become more complex, modular, and configurable supporting mass customization to meet specific customer needs. Hence, Industry 4.0 is characterized by innovation and introduction of new products and services as embedded systems based on knowledge created along the value chain (Fettermann, Gobbo Sá Cavalcante, & Domingues de Almeida Tortorella, 2018).

The servitization strategy usually shows lack of knowledge regarding the service offering associated with their manufactured products; thus, acquiring external knowledge from suppliers and customers can be a way to tackle this problem. The service-oriented approach by means of data and new knowledge leads to new business models (Müller, Buliga, et al., 2018). In this case, the main sources of knowledge are both the new technologies and the interactions with the suppliers and customers and the benefits of data depend on digital and data processes capabilities that are, as for the processes, the main challenge companies need to face (Schroeder et al., 2019).

3.4 Industry 4.0 and Knowledge Management for and from People

Manufacturing companies approaching the Industry 4.0 revolution need to consider the human resources as strategically essential to benefit effectively from new technologies and knowledge they allow to create. The fourth industrial revolution is strategically driven by creative and open-minded people, rather than on technology itself (Ramzi et al., 2018). The new disruptive technologies such as big data, artificial intelligence, and cloud computing are penetrating all manufacturing industries and others bringing together the physical and virtual worlds with higher interconnections inside and beyond the company's boundaries. This new digital scenario needs humans both in terms of interactions with external environment and of upgraded competences to manage technologies, data, and knowledge (Agolla, 2018).

Empirical studies stressed the strategic role of people for the success of Industry 4.0 and for the exploitation of benefits connected to the new technologies. People are considered both as a driver of adoption (Arnold & Voigt, 2019) and as an enabler for the business processes improvements and the achievement of company' strategic goals (Tortorella et al., 2020). Indeed, several studies highlight the digital skills and competences as one of the main challenges of Industry 4.0. High-developed cognitive and processual competencies promote the company's digital transformation (Butschan et al., 2019). Lack of technological expertise (Rauch et al., 2019), qualified employment and qualified engineering staff (Saniuk & Saniuk, 2018), data management skills (Müller, Kiel, et al., 2018; Schroeder et al., 2019), and new technological skills to manage processes and new product development (Krzywdzinski, 2017) are the main critical success factors (Mihardjo et al., 2019; Sivathanu, 2019). In this case, new knowledge is needed to manage digital transformation; thus, companies should support the creation of new skills and competences through different ways, such as training (Mittal et al., 2019) and interactions among works of internal departments (Kohnová et al., 2019).

In addition to organizational and managerial digital practices and capabilities also the horizontal and vertical integrations affect the implementation of Industry 4.0 technologies (Agostini & Filippini, 2019; Bienhaus & Haddud, 2018) facilitating the creation and sharing of knowledge with positive effects on production process and products (Buchi et al., 2020). Data management coming from the integration of new technologies with both company and partner's technological endowment allows improvements of production process (with effects also on backshoring strategies), product quality, servitization, and business model innovation (Dachs et al., 2019; Dalmarco et al., 2019; Jerman et al., 2019).

New technologies have a positive role also in enhancing the customer experience basing on the real-time information sharing (Ghouri & Mani, 2019). Customers are the other essential factor of Industry 4.0. They are precious source of knowledge to personalize products and services (Müller, Buliga, et al., 2018), but also have a key role for the innovation of business model (Kohnová et al., 2019), and thus they are seen as drivers of Industry 4.0 implementation (Moeuf et al., 2020; Müller, 2019a).

People are the most important factor from the viewpoint of knowledge creation and management in the Industry 4.0 paradigm. This new technological approach 4.0 requires changes in talent management practices and the nature of work skill-sets (technical knowledge, networked and collaborative skills, and interdisciplinary skills) (Whysall et al., 2019). The manufacturing digitalization requires qualified workforce and key partners to benefit from operation and marketing opportunities (Arnold & Voigt, 2019; Paruse, 2019).

4 Conclusion

Manufacturing companies approach Industry 4.0 as part of the overall firm strategy (Kane, Palmer, Phillips, Kiron, & Buckley, 2015). Firms adopt new technologies because they expect to improve business processes (operation, marketing, internationalization, value chain, etc.) and sustain their competitive position (Bharadwaj, El Sawy, Pavlou, & Venkatraman, 2013). In the Industry 4.0 paradigm, knowledge assumes a key role as both driver and enabler affecting the decision to implement the Industry 4.0 technologies (Wilkesmann & Wilkesmann, 2018). In this sense, Industry 4.0 is a data-driven paradigm because it enables better use of data produced by technologies and also by interactions along the value chain to create more value (Klingenberg et al., 2019). Not surprisingly, many Industry 4.0 technologies are in some way related to data (named data-driven technologies), which enable successful acquisition of useful data, but also other technologies more related to operation activities depend on data gathered within and outside the firm boundaries. Indeed, the achievements of production and/or marketing outcomes through the digital technologies are linked to the creation of new knowledge for both production and product improvements (Lee, Davari, Singh, & Pandhare, 2018). Thus, the decision-making process in the Industry 4.0 and smart manufacturing systems requires information and knowledge, which can be mined from large amounts of production data (Tao et al., 2018).

The review of empirical studies about the adoption of Industry 4.0 technologies allows us a better understanding of the role of knowledge in the fourth industrial revolution. Firstly, we need to specify that Industry 4.0 is a knowledge-based approach and that knowledge coming from different sources, from the use of new technologies and, thus, from internal processes, but also from the interactions with suppliers and customers. In this case, company should have internally the right digital competences to manage successfully the data gathered, but also to create and manage new knowledge to improve business processes. Digital skills and competences are the most important variable of the relationship between Industry 4.0 and knowledge management because of their strategic role in achieving the business goals (Ferraris et al., 2019) and improving decision process (He, Wang, & Akula, 2017) and firm strategy (Xu, Frankwick, & Ramirez, 2016).

Theoretically, the review of empirical studies allows us to advance the literature on the relationship between Industry 4.0 and knowledge management stressing the holistic role of knowledge in the new digital revolution. The new technologies allow manufacturing companies to improve business processes and customize products and services through data and the new knowledge they are able to generate. The creation of new knowledge depends both on the use of new technologies and on the interactions along the value chain (suppliers and customers). However, the achievement of business benefits strictly depends on human resources and, more specifically, on digital skills and competences. From this perspective, manufacturing companies that approach the Industry 4.0 paradigm should consider such new technologies as new tools that enable new knowledge creation; therefore, they should give attention to the digital skills and competences needed to manage this technological transformation, fostering internal competence upgrading.

The chapter gives some hints about the future directions. Future research should consider the role of firm strategy (in terms of data-driven products/services and human resource) on knowledge management. In particular, more papers (such as conference proceedings and book chapters) from different disciplines would be considered for an overall understanding of the topic.

References

Abubakar, A. M., Elrehail, H., Alatailat, M. A., & Elçi, A. (2019). Knowledge management, decision-making style and organizational performance. *Journal of Innovation & Knowledge, 4*(2), 104–114.

Acharya, A., Singh, S. K., Pereirac, V., & Singh, P. (2018). Big data, knowledge co-creation and decision making in fashion industry. *International Journal of Information Management, 42*, 9–01.

Adamson, G., Wang, L., & Moore, P. (2017). Feature-based control and information framework for adaptive and distributed manufacturing in cyber physical systems. *Journal of Manufacturing Systems, 43*, 305–315.

Agolla, J. E. (2018). Human capital in the smart manufacturing and Industry 4.0 revolution. In A. Petrillo, R. Cioffi, & F. De Felice (Eds.), *Digital Transformation in Smart Manufacturing* (pp. 41–58). London: IntechOpen.

Agrawal, A., Schaefer, S., & Funke, T. (2018). Incorporating Industry 4.0 in corporate strategy. In R. Brunet-Thornton & F. Martinez (Eds.), *Analyzing the Impacts of Industry 4.0 in Modern Business Environments* (pp. 161–176). Hersley, PA: IGI-Global.

Agostini, L., & Filippini, R. (2019). Organizational and managerial challenges in the path toward Industry 4.0. *European Journal of Innovation Management, 22*(3), 406–421.

Agostini, L., & Nosella, A. (2019). The adoption of Industry 4.0 technologies in SMEs: Results of an international study. *Management Decision.* https://doi.org/10.1108/MD-09-2018-0973.

Almada-Lobo, F. (2015). The Industry 4.0 revolution and the future of manufacturing execution systems (MES). *Journal of Innovation Management, 3*(4), 16–21.

Anderson, C. (2012). *Makers: The new industrial revolution.* New York: Crown Business Books.

Ardito, L., Messeni Petruzzelli, A., Panniello, U., & Garavelli, A. C. (2019). Towards Industry 4.0 Mapping digital technologies for supply chain management-marketing integration. *Business Process Management Journal, 25*(2), 323–346.

Ardolino, M., Rapaccini, M., Saccani, N., Gaiardelli, P., Crespi, G., & Ruggeri, C. (2018). The role of digital technologies for the service transformation of industrial companies. *International Journal of Production Research, 56*(6), 2116–2132.

Arnold, C., & Voigt, K.-I. (2019). Determinants of industrial internet of things adoption in German manufacturing companies. *International Journal of Innovation and Technology Management, 16*(6), https://doi.org/10.1142/S021987701950038X.

Arnold, C., Kiel, D., & Voigt, K.-I. (2016). How the industrial internet of things changes business models in different manufacturing industries. *International Journal of Innovation Management, 20*(8), 56. https://doi.org/10.1142/S1363919616400156.

Basl, J. (2018). Pilot study of readiness of Czech companies to implement the principles of Industry 4.0. *Management and Production Engineering Review, 8*(2), 3–8.

Berman, B. (2012). Digital transformation: Opportunities to create new business models. *Strategy & Leadership, 40*(2), 16–24.

Bienhaus, F., & Haddud, A. (2018). Procurement 4.0: Factors influencing the digitisation of procurement and supply chains. *Business Process Management Journal, 24*(4), 965–984.

Bogers, M., Hadar, R., & Bilberg, A. (2016). Additive manufacturing for consumer-centric business models: Implications for supply chains in consumer goods manufacturing. *Technological Forecasting and Social Change, 102*, 225–239.

Bharadwaj, A., El Sawy, O., Pavlou, P., & Venkatraman, N. (2013). Digital business strategy: Toward a next generation of insights. *MIS Quarterly, 37*(2), 471–482.

Büchi, G., Cugno, M., & Castagnoli, R. (2020). Smart factory performance and Industry 4.0. *Technological Forecasting & Social Change, 150*, 119790. https://doi.org/10.1016/j.techfore.2019.119790.

Butschan, J., Heidenreich, S., Weber, B., & Kraemer, T. (2019). Tackling hurdles to digital transformation – the role of competencies for successful industrial internet of things (IIoT) implementation. *International Journal of Innovation Management, 23*(4), 1–34. https://doi.org/10.1142/S1363919619500361.

Byrne, G., Ahearne, E., Cotterell, M., Mullany, B., O'Donnell, G. E., & Sammler, F. (2016). High Performance Cutting (HPC) in the new era of digital manufacturing: A roadmap. *Procedia CIRP, 46*, 1–6.

Cavalcante, I. M., Frazzon, E. M., Forcellini, F. A., & Ivanov, D. (2018). A supervised machine learning approach to data-driven simulation of resilient supplier selection in digital manufacturing. *International Journal of Information Management, 49*, 86–97.

Cepeda Carrión, G., Luis Galán González, J., & Leal, A. (2004). Identifying key knowledge area in the professional services industry: A case study. *Journal of Knowledge Management, 8*(6), 131–150.

Chen, D., Heyer, S., Ibbotson, S., Salonitis, K., Steingrímsson, J. G., & Thiede, S. (2015). Direct digital manufacturing: Definition, evolution, and sustainability implications. *Journal of Cleaner Production, 107*, 615–625.

Chiarello, F., Trivelli, L., Bonaccorsi, A., & Fantoni, G. (2018). Extracting and mapping Industry 4.0 technologies using Wikipedia. *Computers Industry, 100*, 244–257.

da Silva, V. L., Kovaleski, J. L., & Negri Pagani, R. (2019). Technology transfer in the supply chain oriented to Industry 4.0: A literature review. *Technology Analysis & Strategic Management, 31*(5), 546–562.

Dachs, B., Kinkel, S., & Jäger, A. (2019). Bringing it all back home? Backshoring of manufacturing activities and the adoption of Industry 4.0 technologies. *Journal of World Business, 54*(6), 101017. https://doi.org/10.1016/j.jwb.2019.101017.

Dalenogare, L. S., Benitez, B. G., Ayala, N. F., & Frank, A. G. (2018). The expected contribution of Industry 4.0 technologies for industrial performance. *International Journal of Production Economics, 204*, 383–394.

Dalmarco, G., Ramalho, F. R., Barros, A. C., & Soares, A. L. (2019). Providing Industry 4.0 technologies: The case of a production technology cluster. *Journal of High Technology Management Research*. https://doi.org/10.1016/j.hitech.2019.100355.

Di Maria, E., Bettiol, M., Capestro, M., & Furlan, A. (2018). Do Industry 4.0 technologies lead to more (and better) knowledge? In E. Bolisani, E. Di Maria, & E. Scrso (Eds.), *Proceedings of the 19th European Conference on Knowledge Management* (pp. 174–181). Reading: Academic Conferences and Publishing International Ltd.

Dragicevic, N., Ullrich, A., Tsui, E., & Gronau, N. (2019). A conceptual model of knowledge dynamics in the Industry 4.0 smart grid scenario. *Knowledge Management Research & Practice*. https://doi.org/10.1080/14778238.2019.1633893.

Durach, C. F., Wieland, A., & Machuca, J. A. D. (2015). Antecedents and dimensions of supply chain robustness: A systematic literature review. *International Journal of Physical Distribution and Logistics Management, 45*(1/2), 118–137.

Fatorachian, H., & Kazemi, H. (2018). A critical investigation of Industry 4.0 in manufacturing: Theoretical operationalisation framework. *Production Planning & Control, 29*(8), 633–644.

Feng, S. C., Bernstein, W. Z., Hedberg, J. T., & Barnard Feeney, A. (2017). Toward knowledge management for smart manufacturing. *Journal of Computing and Information Science in Engineering, 17*, 031016. https://doi.org/10.1115/1.4037178.

Ferraris, A., Mazzoleni, A., & Devalle, A. (2019). Big data analytics capabilities and knowledge management: Impact on firm performance. *Management Decision, 57*(8), 1923–1936.

Fettermann, D. C., Gobbo Sá Cavalcante, C., Domingues de Almeida, T., & Tortorella, G. L. (2018). How does Industry 4.0 contribute to operations management? *Journal of Industrial and Production Engineering, 35*(4), 255–268.

Frank, A. G., Dalenogare, L. S., & Ayala, N. F. (2019). Industry 4.0 technologies: Implementation patterns in manufacturing companies. *International Journal of Production Economics, 210*, 15–26.

Gilchrist, A. (2016). *Industry 4.0: The industrial Internet of Things*. New York: Springer.

Ghobakhloo, M. (2018). The future of manufacturing industry: A strategic roadmap toward Industry 4.0. *Journal of Manufacturing Technology Management, 29*(6), 910–936.

Ghouri, A. M., & Mani, V. (2019). Role of real-time information-sharing through SaaS: An Industry 4.0 perspective. *International Journal of Information Management, 46*, 301–315.

Hakanen, T., & Jaakkola, E. (2012). Co-creating customer-focused solutions within business networks: A service perspective. *Journal of Service Management, 23*, 593–611.

Haenlein, M., & Kaplan, A. (2019). A brief history of artificial intelligence: On the past, present, and future of artificial intelligence. *California Management Review, 61*(4), 5–14.

He, W., Wang, F. K., & Akula, V. (2017). Managing extracted knowledge from big social media data for business decision making. *Journal of Knowledge Management, 21*(2), 275–294.

Holmström, J., Holweg, M., Khajavi, S. H., & Partanen, J. (2016). The direct digital manufacturing (r)evolution: definition of a research agenda. *Operations Management Research, 9*(1-2), 1–10.

Hozdić, E. (2015). Smart factory for Industry 4.0: A review. *International Journal of Modern Manufacturing Technology, 2*(1), 2067–3604.

Jerman, A., Erenda, I., & Bertoncelj, A. (2019). The influence of critical factors on business model at a smart factory: A case study. *Business Systems Research, 10*(1), 42–52.

Jung, K., Choi, S., Kulvatunyou, B., Cho, H., & Morris, K. (2017). A reference activity model for smart factory design and improvement. *Production Planning & Control, 28*(2), 108–122.

Kagermann, H. (2015). Change through digitization–Value creation in the age of Industry 4.0. In H. Albach, H. Meffert, A. Pinkwart, & R. Reichwald (Eds.), *Management of Permanent Change* (pp. 23–45). Wiesbaden: Springer.

Kagermann, H., Helbig, J., & Wahlster, W. (2013). *Recommendations for implementing the strategic initiative Industrie 4.0*. (Final report of the Industrie 4.0 Working Group) (pp. 1–84). Berlin: Forschungsunion.

Kagermann, H., Lukas, W.-D., & Wahlster, W. (2011). *Industrie 4.0: Mit dem internet der dinge auf dem weg zur 4. industriellen revolution* (p. 13). Düsseldorf: VDI Nachrichten.

Kagermann, H., Gausemeier, A. J., Schuh, G., & Wahlster, W. (2016). *Industrie 4.0 in a Global Context*. Munich: Herbert Utz Verlag.

Kamble, S., Gunasekaran, A., & Dhone, N. C. (2020). Industry 4.0 and lean manufacturing practices for sustainable organisational performance in Indian manufacturing companies. *International Journal of Production Research, 58*, 1319–1337.

Kane, G., Palmer, D., Phillips, A., Kiron, D., & Buckley, N. (2015). Strategy, not technology, drives digital transformation. *MIT Sloan Management Review, 14*, 1–25.

Karia, N. (2018). Knowledge resources, technology resources and competitive advantage of logistics service providers. *Knowledge Management Research & Practice, 16*(3), 414–426.

Kiel, D., Arnold, C., & Voigt, K.-I. (2017). The influence of the industrial Internet of Things on business models of established manufacturing companies – A business level perspective. *Technovation, 68*, 4–19.

Kiel, D., Müller, J. M., Arnold, C., & Voigt, K.-I. (2017). Sustainable industrial value creation: Benefits and challenges of Industry 4.0. *International Journal of Innovation Management, 21*(8), 259–280.

Klingenberg, C. O., Viana Borges, M. A., & Valle Antunes Jr., J. A. (2019). Industry 4.0 as a data-driven paradigm: a systematic literature review on technologies. *Journal of Manufacturing Technology Management.* https://doi.org/10.1108/JMTM-09-2018-0325.

Kohler, D., & Weisz, J.-D. (2016). *Industrie: Les défis de la transformation numérique du modèle industriel allemand.* Paris: La documentation Française.

Kohnová, L., Papula, J., & Salajova, N. (2019). Internal factors supporting business and technological transformation in the context of Industry 4.0. *Business: Theory and Practice, 20*, 137–145.

Krzywdzinski, M. (2017). Automation, skill requirements and labour-use strategies: High-wage and low-wage approaches to high-tech manufacturing in the automotive industry. *New Technology, Work and Employment, 32*(3), 247–267.

Kusiak, A. (2018). Smart manufacturing. *International Journal of Production Research, 56*(1–2), 508–517.

Liao, Y., Deschamps, F., Loures, E. d. F. R., & Ramos, L. F. P. (2017). Past, present and future of Industry 4.0 – A systematic literature review and research agenda proposal. *International Journal of Production Research, 55*(12), 3609–3629.

Lasi, H., Kemper, H. G., Fettke, P., Feld, T., & Hoffmann, M. (2014). Industry 4.0. *Business & Information Systems Engineering, 6*(4), 239–242.

Lee, J., Davari, H., Singh, J., & Pandhare, V. (2018). Industrial artificial intelligence for Industry 4.0-based manufacturing systems. *Manufacturing Letters, 18*, 20–23.

Lee, J., Bagheri, B., & Kao, H. A. (2015). A cyber-physical systems architecture for Industry 4.0-based manufacturing systems. *Manufacturing Letters, 3*, 18–23.

Lin, D., Lee, C. K. M., Lau, H., & Yang, Y. (2018). Strategic response to Industry 4.0: an empirical investigation on the Chinese automotive industry. *Industrial Management and Data Systems, 118*(3), 589–605.

Lu, Y. (2017). Industry 4.0: A survey on technologies, applications and open research issues. *Journal of Industrial Information Integration, 6*, 1–10.

Lu, H.-P., & Weng, C.-I. (2018). Smart manufacturing technology, market maturity analysis and technology roadmap in the computer and electronic product manufacturing industry. *Technological Forecasting & Social Change, 133*, 85–94.

Lugert, A., Batz, A., & Winkler, H. (2018). Empirical assessment of the future adequacy of value stream mapping in manufacturing industries. *Journal of Manufacturing Technology Management, 29*(5), 886–906.

Mihardjo, L. W. W., Sasmoko, S., Alamsjah, F., & Elidjen, E. (2019). Digital leadership role in developing business model innovation and customer experience orientation in Industry 4.0. *Management Science Letters, 9*(11), 1749–1762.

Mittal, S., Khan, M. A., Purohit, J. K., Menon, K., Romero, D., & Wuest, T. (2019). A smart manufacturing adoption framework for SMEs. *International Journal of Production Research.* https://doi.org/10.1080/00207543.2019.1661540.

Mittal, S., Khan, M. A., Romero, D., & Wuest, T. (2018). A critical review of smart manufacturing & Industry 4.0 maturity models: Implications for small and medium-sized enterprises (SMEs). *Journal of Manufacturing Systems, 49*, 194–214.

Moeuf, A., Lamouri, S., Pellerin, R., Tamayo-Giraldo, S., Tobon-Valencia, E., & Eburdy, R. (2020). Identification of critical success factors, risks and opportunities of Industry 4.0 in SMEs. *International Journal of Production Research, 58*, 1384–1400. https://doi.org/10.1080/00207543.2019.1636323.

Moeuf, A., Pellerin, R., Lamouri, S., Tamayo-Giraldo, S., & Barbaray, R. (2018). The industrial management of SMEs in the era of Industry 4.0. *International Journal of Production Research, 56*(3), 1118–1136.

Mohamed, M. (2018). Challenges and benefits of Industry 4.0: An overview. *International Journal of Supply and Operations Management., 5*(3), 256–265.

Müller, J. M. (2019a). Business model innovation in small- and medium-sized enterprises: Strategies for Industry 4.0 providers and users. *Journal of Manufacturing Technology Management.* https://doi.org/10.1108/JMTM-01-2018-0008.

Müller, J. M. (2019b). Antecedents to digital platform usage in Industry 4.0 by established manufacturers. *Sustainability, 11*, 1121. https://doi.org/10.3390/su11041121.

Müller, J. M., & Voigt, K. I. (2018). Sustainable industrial value creation in SMEs: A comparison between Industry 4.0 and Made in China 2025. *International Journal of Precision Engineering and Manufacturing – Green Technology, 5*(5), 659–670.

Müller, J. M., Buliga, O., & Voigt, K.-I. (2018). Fortune favors the prepared: How SMEs approach business model innovations in Industry 4.0. *Technological Forecasting & Social Change, 132*, 2–17.

Müller, J. M., Kiel, D., & Voigt, K.-I. (2018). What drives the implementation of Industry 4.0? The role of opportunities and challenges in the context of sustainability. *Sustainability, 10*, 247. https://doi.org/10.3390/su10010247.

Nagy, J., Oláh, J., Erdei, E., Máté, D., & Popp, J. (2018). The role and impact of Industry 4.0 and the Internet of Things on the business strategy of the value chain—The case of Hungary. *Sustainability, 10*, 3491. https://doi.org/10.3390/su10103491.

Neirotti, P., Raguseo, E., & Paolucci, E. (2018). How SMEs develop ICT-based capabilities in response to their environment: Past evidence and implications for the uptake of the new ICT paradigm. *Journal of Enterprise Information Management, 31*(1), 10–37.

Panetto, H., & Molina, A. (2008). Enterprise integration and interoperability in manufacturing systems: Trends and issues. *Computers in Industry, 59*(7), 641–646.

Porter, M., & Heppelmann, J. (2015). How smart, connected products are transforming companies. *Harvard Business Review, 93*(10), 1–19.

Porter, M. E., & Heppelmann, J. E. (2014). How smart, connected products are transforming competition. *Harvard Business Review, 92*(11), 64–88.

Prause, M. (2019). Challenges of Industry 4.0 technology adoption for SMEs: The case of Japan. *Sustainability, 11*, 5807. https://doi.org/10.3390/su11205807.

Prause, G., & Atari, S. (2017). On sustainable production networks for Industry 4.0. *The International Journal Entrepreneurship and Sustainability Issues, 4*(4), 421–431.

Rajput, S., & Singh, S. P. (2019). Connecting circular economy and Industry 4.0. *International Journal of Information Management, 49*, 98–113.

Ramzi, B., Ahmad, H., & Zakaria, N. (2018). A conceptual model on people approach and smart manufacturing. *International Journal of Supply Chain Management, 8*(4), 1102–1107.

Rashid, A., & Tjahjono, B. (2016). Achieving manufacturing excellence through the integration of enterprise systems and simulation. *Production Planning & Control, 27*(10), 837–852.

Rauch, E., Dallasega, P., & Unterhofer, M. (2019). Requirements and barriers for introducing smart manufacturing in small and medium-sized enterprises. *IEEE Engineering Management Review, 47*(3), 87–94.

Reinhard, G., Jesper, V. & Stefan, S. (2016). *Industry 4.0: Building the digital enterprise* (pp. 1–39). 2016 Global Industry 4.0 Survey.

Reischauer, G. (2018). Industry 4.0 as policy-driven discourse to institutionalize innovation systems in manufacturing. *Technological Forecasting and Social Change, 132*, 26–33.

Roblek, V., Meško, M., & Krapež, A. (2016). A complex view of Industry 4.0. *SAGE Open, 6*(2), 1–11.

Rojko, A. (2017). Industry 4.0 concept: Background and overview. *International Journal of Interactive Mobile Technologies, 11*(5), 77–90.

Saniuk, S., & Saniuk, A. (2018). Challenges of Industry 4.0 for production enterprises functioning within cyber industry networks. *Management Systems in Production Engineering, 26*(4), 212–216.

Santos, M. Y., e Sá, J. O., Andrade, C., Lima, F. V., Costa, E., Costa, C., Martinho, B., & Galvão, J. (2017). A Big Data system supporting Bosch Braga Industry 4.0 strategy. *International Journal of Information Management, 37*(6), 750–760.

Schlechtendahl, J., Keinert, M., Kretschmer, F., Lechler, A., & Verl, A. (2015). Making existing production systems Industry 4.0-ready. *Production Engineering, 9*(1), 143–148.

Schmidt, R., Möhring, M., Härting, R.-C., Reichstein, C., Neumaier, P., & Jozinović, P. (2015). *Industry 4.0- potentials for creating smart products: Empirical research results.* Heidelberg: Springer.

Schneider, P. (2018). Managerial challenges of Industry 4.0: An empirically backed research agenda for a nascent field. *Review of Managerial Science, 12*(3), 803–848.

Schniederjans, D.,G., Curado, C., & Khalajhedayati, M. (2019). Supply chain digitisation trends: An integration of knowledge management. *International Journal of Production Economics*. https://doi.org/10.1016/j.ijpe.2019.07.012.

Schroeder, A., Ziaee Bigdeli, A., Galera Zarco, C., & Baines, T. (2019). Capturing the benefits of Industry 4.0: a business network perspective. *Production Planning and Control, 30*(16), 1305–1321.

Schwab, K. (2017). *The fourth industrial revolution.* New York: Crown.

Seetharaman, A., Patwa, N., Saravanan, A. S., & Sharma, A. (2019). Customer expectation from Industrial Internet of Things (IIOT). *Journal of Manufacturing Technology Management.* https://doi.org/10.1108/JMTM-08-2018-0278.

Sivathanu, B. (2019). Adoption of industrial IoT (IIoT) in auto-component manufacturing SMEs in India. *Information Resources Management Journal, 32*(2), 52–75.

Ślusarczyk, B., Haseeb, M., & Hussain, H. I. (2019). Fourth industrial revolution: A way forward to attain better performance in the textile industry. *Engineering Management in Production and Services, 11*(2), 52–69.

Stadnicka, D., Litwin, P., & Antonelli, D. (2019). Human factor in intelligent manufacturing systems – knowledge acquisition and motivation. *Proceedia CIRP, 79*, 718–723.

Stentoft, J. and Rajkumar, C. (2019). The relevance of Industry 4.0 and its relationship with moving manufacturing out, back and staying at home. *International Journal of Production Research.* https://doi.org/10.1080/00207543.2019.1660823.

Szalavetz, A. (2019). Industry 4.0 and capability development in manufacturing subsidiaries. *Technological Forecasting & Social Change, 145*, 384–395.

Tao, F., Qi, Q., Liu, A., & Kusiak, A. (2018). Data-driven smart manufacturing. *Journal of Manufacturing Systems, 48*, 157–169.

Thoben, K.-D., Wiesner, S., & Wuest, T. (2017). "Industrie 4.0" and smart manufacturing - A review of research issues and application examples. *International Journal of Automation Technology, 11*(1), 4–16.

Tortorella, G. L., & Fettermann, D. (2018). Implementation of Industry 4.0 and lean production in Brazilian manufacturing companies. *International Journal of Production Research, 56*(8), 2975–2987.

Tortorella, G. L., Msc Cawley Vergara, A., Garza-Reyesc, J. A., & Sawhneyd, R. (2020). Organizational learning paths based upon industry 4.0 adoption: An empirical study with Brazilian manufacturers. *International Journal of Production Economics, 219*, 284–294.

Tranfield, D., Denyer, D., & Smart, P. (2003). Towards a methodology for developing evidence-informed management knowledge by means of systematic review. *British Journal of Management, 14*(3), 207–222.

Wagire, A. A., Rathore, A. P. S., & Rakesh, J. (2019). Analysis and synthesis of Industry 4.0 research landscape. Using latent semantic analysis approach. *Journal of Manufacturing Technology Management, 31*, 31–51. https://doi.org/10.1108/JMTM-10-2018-0349.

Wang, Y., Hai-Shu Ma, H.-S., Yang, J.-H., & Wang, K.-S. (2017). Industry 4.0: a way from mass customization to mass personalization production. *Advances in Manufacturing, 5*(4), 311–320.

Wee, D., Kelly, R., Cattel, J., & Breunig, M. (2015). *Industry 4.0—How to navigate digitization of the manufacturing sector.* Munich: McKinsey & Company.

Whysall, Z., Owtram, M., & Brittain, S. (2019). The new talent management challenges of Industry 4.0. *Journal of Management Development, 38*(2), 118–129.

Wilkesmann, M., & Wilkesmann, U. (2018). Industry 4.0 – organizing routines or innovations? *VINE Journal of Information and Knowledge Management Systems, 48*(2), 238–254.

Xu, Z., Frankwick, G. L., & Ramirez, E. (2016). Effects of big data analytics and traditional marketing analytics on new product success: A knowledge fusion perspective. *Journal of Business Research, 69*(5), 1562–1566.

Yin, Y., Stecke, K. E., & Li, D. (2017). The evolution of production systems from industry 2.0 through Industry 4.0. *International Journal of Production Research, 56*(1-2), 848–861.

Zhong, R. Y., Xu, C., Chen, C., & Huang, G. Q. (2017). Big Data Analytics for physical Internet-based Intelligent Manufacturing shop floors. *International Journal of Production Research, 55*(9), 2610–2621.

Zhong, R. Y., Xu, X., Klotz, E., & Newman, S. T. (2017). Intelligent manufacturing in the context of Industry 4.0: A review. *Engineering, 3*, 613–630.

Do Industry 4.0 Technologies Matter When Companies Backshore Manufacturing Activities? An Explorative Study Comparing Europe and the US

Luciano Fratocchi and Cristina Di Stefano

Abstract The objective of this chapter is to analyze the impact (if any) of Industry 4.0 enabling technology on firms' decision to relocate to the home country their offshored production activities. In particular, the chapter analyzes whether Industry 4.0 technologies may represent a driver/motivation or an enabling factor for companies which are evaluating such a strategic alternative. In order to reach such an objective, a two-step explorative methodology has been applied. After implementing a structured literature review, empirical evidence of backshoring decisions implemented by both European and US companies has been analyzed. Collected findings show that the majority of sampled articles conceptualize Industry 4.0 technologies as a driver. At the same time, empirical findings show some interesting differences between European and US companies adopting backshoring decisions based on/enabled by Industry 4.0 technologies. Finally, competences (related to both the manufacturing activities as a whole and the Industry 4.0 technologies) emerge as one of the most critical issue for investigated companies.

1 Introduction

Companies have been offshoring (and often also outsourcing) their manufacturing activities for a long time. They mostly relocate to low-cost countries (e.g., Eastern Europe and Asia) since their main goal was efficiency seeking. However, the benefits of offshoring have often proven elusive (Manning, 2014); for instance, the relocation of production activities abroad often diminishes firm's competence due to the spatial decoupling of R&D and manufacturing activities (Stentoft, Olhager, Heikkilä, & Thoms, 2016). This risk is even higher when offshoring decisions are coupled with the adoption of outsourcing governance mode. In such a context, employee deskilling and decline of firms' industrial knowledge emerge (Nujen,

---

L. Fratocchi (✉) · C. Di Stefano
University of L'Aquila, L'Aquila, Italy
e-mail: luciano.fratocchi@univaq.it

Halse, Damm, & Gammelsaeter, 2018). This, in turn, may have serious implications also for the entire economic system of the home country level (Pisano & Shih, 2012). Therefore, in the last 20 years an increasing number of companies have been reconsidering their offshoring choice having experienced several offshoring difficulties (Manning, 2014). Consequently, they often adopt a relocation of second-degree strategy (Barbieri, Elia, Fratocchi, & Golini, 2019), also identified by the literature as reshoring (Fratocchi, Di Mauro, Barbieri, Nassimbenid, & Zanoni, 2014). This term includes both the relocation to the home country (RHC or backshoring) and the one to a third country (RTC). The latter is defined alternatively near-shoring—when the company relocates to a host country within the home region—and further offshoring—when the new host country is a faraway one.

In the last 10 years, scholars have mostly focused their attention on RHC operations (Barbieri, Ciabuschi, Fratocchi, & Vignoli, 2018; Stentoft et al., 2016; Wiesmann, Snoei, Hilletofth, & Eriksson, 2017) particularly studying motivations, i.e., drivers of the operations. Among them, increasing attention has been paid to production automation (see, for instance, Ancarani & Di Mauro, 2018; Ancarani, Di Mauro, & Mascali, 2019) and additive manufacturing (Fratocchi, 2018a, 2018b; Moradlou & Tate, 2018). Both of them are technologies based on cyber-physical systems, and are identified with the broad term Industry 4.0 technologies i.e., "smart machines, warehousing systems and production facilities that have been developed digitally and feature end-to-end ICT-based integration, from inbound logistics to production, marketing, outbound logistics and service" (Kagermann, Wahlster, & Helbig, 2013, see p. 14).

Firm's internationalization process can be strongly influenced by information and communications technologies (ICTs); they allow remote coordination and extend the span of control while reducing its cost (Alcácer, Cantwell, & Piscitello, 2016; Chen & Kamal, 2016; Leamer & Storpe, 2001). Thanks to those technologies, companies can redefine their location strategy and "fine slice" the most value adding activities (Buckley, 2011; Buckley & Ghauri, 2004) or reconfigure their production footprint. Moreover, the increase in productivity these technologies allow (Brynjolfsson & McAfee, 2014; Kagermann et al., 2013) may reduce—and even eliminate—location advantages of low-cost countries (Ancarani et al., 2019; Ancarani & Di Mauro, 2018; Dachs, Kinkel, & Jäger, 2019). At the same time, the adoption of Industry 4.0 technologies allows a higher flexibility of the manufacturing process and increases companies' responsiveness to clients' need and their possibility to offer customized products (Ancarani et al., 2019; Ancarani & Di Mauro, 2018; Dachs et al., 2019; Fratocchi, 2018a, b; Moradlou, Backhouse, & Ranganathan, 2017; Moradlou & Tate, 2018). Finally, Lampón and González-Benito (2019) have recently showed that companies which improved their key manufacturing resources (e.g., process optimization, technologies, and facilities) after the offshoring decision are more likely to backshore.

At the same time, the implementation of Industry 4.0 technologies requests companies to develop specific competencies (Nujen, Mwesiumo, Solli-Sæther, & Slyngstad, 2018). In this respect, recent studies pointed out that there is a serious lack of qualified workforce able to implement such technologies, especially in small and medium companies (Stentoft, Jensen, Philipsen, & Haug, 2019; Stentoft & Rajkumar, 2019). Therefore, companies aiming at implementing backshoring strategies need to evaluate their readiness not only in terms of manufacturing competences (Lampón & González-Benito, 2019) but also in terms of Industry 4.0 ones (Nujen, Mwesiumo, et al., 2018).

Considering the above-discussed framework, this chapter mainly addresses two research questions:

(a) In the evaluation of RSD alternatives, do companies consider Industry 4.0 technologies as a driver/motivation (Fratocchi et al., 2016)?
(b) In the evaluation of RSD alternatives, do companies consider Industry 4.0 technologies as an enabling factor (Engström, Hilletofth, Eriksson, & Hilletofth, 2018; Engström, Sollander, Hilletofth, & Eriksson, 2018)?

A two-step explorative approach will be adopted to investigate the two research questions, the first of which is conducted through a structured literature review based on 115 Elsevier Scopus indexed journal articles published until August 2019. The second step of the adopted methodology is based on empirical evidence based on the UnivAQ Manufacturing Reshoring Dataset (UMRD), which has already been adopted in previous backshoring research (Ancarani et al., 2019; Ancarani & Di Mauro, 2018; Ancarani, Di Mauro, Fratocchi, Orzesc, & Sartorc, 2015; Fratocchi, 2018a, b; Fratocchi et al., 2015; Fratocchi et al., 2016; Wan, Orzes, Sartor, & Nassimbeni, 2019; Wan, Orzes, Sartor, Di Mauro, & Nassimbeni 2019) since it is recognized as the most comprehensive at the worldwide level.

The first step of the analysis indicates that interest of scholars in the topic under investigation has been growing over the years. However, among all the Industry 4.0 enabling technologies, the literature has mainly focused on the study of production automation (42 out of 115 sampled Elsevier Scopus indexed journal articles, published from 2014 to 2019) and additive manufacturing (10 documents published only in the last 2 years). Moreover, only four journal articles (of which three have been published in 2019) specifically investigated the causality (if any) of Industry 4.0 technologies on backshoring. However, the research findings emerging from these four articles are quite differentiated and not definitive. Finally, it is worth noting that, while the majority of sampled articles conceptualize Industry 4.0 technologies as a driver (Barbieri et al., 2018), they have also been viewed as enabling factors (Engström, Hilletofth, et al., 2018; Engström, Sollander, et al., 2018). At the same time, empirical findings sorted by the UMRD show some interesting differences between European and US companies adopting backshoring decisions based on/enabled by Industry 4.0 technologies.

To investigate the proposed research questions, the rest of the chapter is as follows: Section 2 describes the methodology adopted. Section 3 presents and discusses findings. The last section concludes and presents the implications and limitations of the analysis.

2 Methodology

As previously introduced, the analysis is conducted adopting a two-step explorative methodology. At first, a structured literature review regarding backshoring decision has been conducted following the Seuring and Gold (2012) approach for content analysis. This approach has been already followed for literature reviews focused on RHC (Barbieri et al., 2018; Stentoft et al., 2016). Documents have been extracted from the Elsevier Scopus dataset, which is recognized as one of the most valuable source for publications in the business and management field of study (Greenwood, 2011). Adopted research criteria were the following:

(a) English written journal articles
(b) Published until August 2019
(c) Containing in the title, abstract, and/or keywords one of the following terms: "reshor∗," "re-shor∗," "backshore∗," "back-shor∗," "back-reshor∗," and "back-sourc∗"

Authors found a total number of 177 journal articles and carefully read all the text. Some articles were excluded from the analysis on the basis of the following excluding criteria:

- Journal articles focusing on RHC implemented by service companies (e.g., ICT companies)
- Not peer review articles
- Journal articles related to different fields of study (reshoring concept is used with different meanings in the maritime and building engineering research fields)
- Journal articles not focused on manufacturing (e.g., documents referring to functions as human resources and research and development (R&D)).
- Based on these criteria, 62 documents were eliminated; therefore, the total amount of sampled documents was 115 (see Appendix).

The second step of the analysis considers the evidence collected in the UMRD; it contains data of European and American companies that implemented RHM operations. To the best of our knowledge, it is the most comprehensive available dataset on reshoring since it combines evidence from different sources:

(a) European Reshoring Monitor (ERM) dataset: it is a public available dataset that has already been used in previous backshoring studies (Ancarani et al., 2019; Wan, Orzes, Sartor, Di Mauro, et al., 2019; Wan, Orzes, Sartor, & Nassimbeni, 2019). It was financed by the EU foundation Eurofound and "collects information on individual reshoring cases from several sources (media, specialized press, scientific literature, practitioner literature) and it organizes it into a secured access, regularly updated, online database" (https://reshoring.eurofound.europa.eu/).
(b) Uni-CLUB MoRe reshoring (UCMR) dataset: it is a vast dataset containing evidence of companies that implemented manufacturing backshoring operations and has already been considered in several researches on manufacturing

backshoring (see, among others, Ancarani & Di Mauro, 2018; Ancarani et al., 2015, 2019; Fratocchi, 2018a, b; Fratocchi et al., 2015, 2016; Wan, Orzes, Sartor, Di Mauro, et al., 2019; Wan, Orzes, Sartor, & Nassimbeni, 2019).

(c) Reshoring Initiative dataset: it is a large dataset which includes evidence of US companies that implemented various location strategies (e.g., backshoring, kept from offshoring, foreign direct investments) having an impact on employment levels in the USA. It was already used for previous research on the phenomenon in the USA (Abbasi, 2016; Moore, Rothenberg, & Moser, 2018). Given the heterogeneity of the operations it includes, all the evidence has been checked by researchers and only the ones referring to RHC decisions have been incorporated in the UMRD dataset.

Up to the end of December 2018, the UMR dataset contained a total of 1279 instances of evidence regarding backshoring decisions implemented by companies belonging to 24 European countries (814), the USA (428), and other foreign countries (37).

3 Findings

3.1 Findings from the Extant Literature

The analysis of the 115 sampled journal articles clearly shows that the relationship (if any) between Industry 4.0 technologies as a whole and backshoring has been specifically addressed by only four journal articles (namely, Ancarani & Di Mauro, 2018; Ancarani et al., 2019; Dachs et al., 2019, Stentoft & Rajkumar, 2018). However, wider attention has been given to two of the most well-known Industry 4.0 technologies, namely automation and three-dimensional (3D) printing/additive manufacturing (Table 1). More specifically, reshoring scholars have been increasingly conceptualizing automation as a backshoring driver and/or an enabling factor since 2014, reaching a total of 42 citations up to August 2019. In contrast, attention to the role of additive manufacturing/3D printing technologies has arisen only in the last 2 years. This finding may be—at least partially—explained by the early stage of the additive manufacturing technologies in large-scale production (Fratocchi, 2018a, b). Finally, only one contribution (Ancarani & Di Mauro, 2018) specifically refers to other two Industry 4.0 technologies, namely sensors and simulation. At the same time, Ancarani et al. (2019) investigated the opportunity for adopting cyber-physical systems to connect production and development and/or buyers and suppliers. Finally, it must be taken into account that the influence (if any) of Industry 4.0 technologies on backshoring decisions has been increasingly proposed as a future research avenue (e.g., Bals, Kirchoff, & Foerstl, 2016; Barbieri et al., 2018; Engström, Hilletofth, et al., 2018; Stentoft et al., 2016). Therefore, this chapter appears to be timely since it allows us to define the state of the art of the academic

Table 1 Breakdown of sample articles by year and article content

Year	Published articles	Production automation	3D printing/additive manufacturing	Sensors	Simulation	Cyber-physical systems connecting production and development and/or buyers and suppliers	I4.0 impact on backshoring
2007	1						
2009	1						
2011	1						
2012	1						
2013	6						
2014	11	3					
2015	8	1					
2016	23	12					
2017	16	6					
2018	24	13	8			1	1
2019	23	7	2	1	1		3
Total	115	42	10	1	1	1	4

debate on Industry 4.0 technologies and second-degree relocations to the home country.

Regarding production automation technology, the first evidence in the sampled journal articles is proposed by Arlbjørn and Mikkelsen (2014) who found that 47.5% of Danish firms which offshored production activities between 2009 and 2014 found the same activities could be backshored as a result of the advances in automation. Similarly, Heikkilä et al. (2018, b) found in a sample of Danish, Finnish, and Swedish companies that access to technology (including production automation) is one of the "significantly more important drivers for back-shoring than for off-shoring ($p \leq 0.001$)" (Heikkilä, Martinsuo, & Nenonen, 2018, p. 228). Moreover, Johansson and Olhager (2018a, b) found, on the basis of a Swedish sample, that companies that have both off- and backshored during the investigated period considered the access to technology at a slightly lower level than companies implementing only backshoring strategies. Finally, in their qualitative study, Engström, Sollander, et al. (2018) found that several companies decided to backshore in Sweden following the benefits offered by production automation. However, the huge contribution of such an enabling technology to the relocation of manufacturing activities in the home Nordic countries seems to be questioned by scholars who investigated other geographic areas. For instance, Ancarani and Di Mauro (2018) point out that only 13.6% of the 840 backshoring decisions belonging to the EU and US companies they analyzed specifically declared at least one of the Industry 4.0 technologies as a relocation driver. At the same time, De Backer, DeStefano, Menon, and Suh (2018) found that robotics have a negative impact on offshoring decisions (at least for companies located in developed countries) but do not yet trigger backshoring decisions.

It has been speculated that production automation reduces the relevance of labor cost as a location criterion since it increases productivity (Abbasi, 2016), making production in high-cost countries more viable (Engström, Hilletofth, et al., 2018). As a consequence, such a production technology has usually been considered as a driver of RHC. It also facilitates the implementation of a flexible production system (Lu, 2017) that allows product customization and firms' responsiveness (Moradlou et al., 2017). Based on this, Ancarani and Di Mauro (2018) state that both "cost-oriented" (i.e., relocation aimed at reducing production and logistics costs) and the "flexibility-oriented" (aimed at improving a firm's responsiveness to customer needs) backshoring strategies are supported by production automation. This evidence is quite relevant since—according to these two authors—the two typologies of reshoring decisions are the most diffused among the 840 backreshoring initiatives' evidence at the worldwide level they analyzed. In contrast, "quality-oriented" backshoring strategies—i.e., when the relocation to the home country is aimed at implementing product upgrade strategies (Bettiol, Burlina, Chiarvesio, & Di Maria, 2018)—are less relevant. This finding is quite at odds with previous evidence collected by Moradlou et al. (2017) and Moradlou and Tate (2018) with respect to the UK backshoring firms. This divergence may, at least partially, be explained from a home country perspective, that is the amount of product and process knowledge located at the home location, either within the backshoring company or within its suppliers' network. In this respect, the relocation within an industrial district at the home country could be not coupled with investments in Industry 4.0 since the backshoring company may implement upgrade strategies leveraging on specific manufacturing competencies (often having craft/manual nature) developed at the cluster level. On the contrary, firms located in countries where manufacturing manual competences are no longer available (given the de-industrialization processes following decades of offshoring strategies) may substitute them with production automation systems (Ancarani & Di Mauro, 2018).

As far as the second research question (Industry 4.0 as a barrier to backshoring strategies) is concerned, Engström, Hilletofth, et al. (2018) are the only authors addressing this issue. More specifically, they point out that Industry 4.0 may represent not only a driver of backshoring decision but also a barrier to its implementation. In this respect, useful insights have been recently offered by Stentoft and Rajkumar (2019). Authors point out that companies characterized by high levels of Industry 4.0 relevance (that is they carefully analyzed drivers and barriers of this phenomenon) are the ones that either backshored or simultaneously off- and backshored in the last 3 years. On the contrary, companies remained at the home country did not develop a specific Industry 4.0 competence. According to the authors, the former companies (the ones backshored or off- and backshored) have been developed or are still involved in learning processes. More specifically, such learning processes might or might not include learning about Industry 4.0 issues. : *"if the level of automation should be seen as a factor acting as a barrier or driver,"* i.e., if it either boosts the backshoring decision or its lack hinders the relocation to the home country. In this respect, Nujen, Halse, et al. (2018) point out that the introduction of new technologies requests new competences within the company; therefore, the implementation of Industry 4.0 programs should be carefully evaluated in terms of firm's backshoring readiness (Bals et al., 2016; Nujen, Mwesiumo, et al., 2018). In this respect, employee upskilling programs are of crucial relevance.

As far as the 3D/additive manufacturing technologies are concerned, it is expected they will have a disruptive impact on global value chains (GVC), therefore also supporting backshoring decisions (Brennan, Ferdows, Godsell, & Golini, 2015; Strange & Zucchella, 2018). In this respect is worth noting that Moradlou and Tate (2018) found that 72% of 50 investigated companies adopting additive manufacturing technologies positively evaluate the contribution it makes to backshoring decisions. In this respect, d'Aveni (2015) states that 3D printing technologies will induce firms to locate manufacturing activities closer to customers; hence its adoption would boost reshoring decisions. Ancarani and Di Mauro (2018) adopt a more restrictive position, stating that this technology may support the implementation of only quality-oriented backshoring decisions. This is because additive manufacturing better supports product development processes and integration between R&D, design, production, and marketing functions (Ketokivi, Turkulainen, Seppälä, Rouvinend, & Ali-Yrkköd, 2017). Moreover, additive manufacturing allows firms to reduce prototyping costs and times (Ancarani et al., 2019). Moreover, Moradlou and Tate (2018) state that relocation to the home country is boosted by the following six benefits that additive manufacturing technologies offer in terms of supply chain management: *"shorter lead time, responsiveness to the product and market changes, lower transportation costs, fewer miscommunications with suppliers, more customization options, fewer products stored in inventory"* (see p. 241). At the same time, Fratocchi (2018a, b) presents evidence that 3D printing technology produces technical and economic advantages that adequately respond to the backshoring drivers presented by the literature (Barbieri et al. 2018). Moreover, Fratocchi (2018a, b) showed that additive manufacturing technologies are adopted in the same industries in which the literature identified greater evidence of backshoring decisions. This is in line with Laplume, Petersen, and Pearce (2016) who identified industries more likely to introduce additive manufacturing technologies.

As already noted, the attention paid by scholars to the relationship (if any)—and even the causality—between manufacturing reshoring and the whole set of Industry 4.0 technologies is still in its infancy. Among the few authors who have investigated such a linkage, Ancarani and Di Mauro (2018) point out "robotics is not a necessary ingredient of [back-]reshoring" but "Industry 4.0 supports manufacturing [back-]reshoring when design and product innovation are involved" (2018, see p. 8). At the same time, Ancarani et al. (2019) provide evidence that—at least until now—backshoring decisions have been implemented without investing in new technologies, especially if the relocation was aimed at leveraging on the "made in" effect and/or shortening the lead time and improving firms' responsiveness. However, authors expect Industry 4.0 may play—in the near future—a specific role in supporting manufacturing relocation decisions, especially in the case of skill shortage—due to previous de-industrialization emerging after decades of manufacturing offshoring—and/or when companies aim to improve design and strengthen product-development linkage. Previous findings are also confirmed by Stentoft and Rajkumar (2019) who analyzed a sample of Danish manufacturing companies. They found that the investigated technologies have no impact on the decision to relocate manufacturing activities to the home country. In contrast, Dachs et al. (2019) found a positive and significant association between investments in Industry 4.0 technologies and backshoring decisions. Moreover, their study—which has been focused on manufacturing companies belonging to Germany, Austria, and Switzerland—also shows that there is no causality between the two variables since both of them are driven by the research on higher levels of flexibility. It is worth noting that a previous investigation on a German sample conducted by Müller, Dotzauer, and Voigt (2017) (not included in the sampled literature) found that in only 13 of the 50 sampled backshoring decisions they analyzed, have Industry 4.0 technologies played a supporting role. Moreover, quantitative analysis of the issue did not support the correlation: considering a Likert scale (from 1 to 5), the mean value was 2.3, for companies that implemented backshoring while in-sourcing their production activities, and 2.2 for those which backshored while outsourcing. Findings by Müller et al. (2017) also show that the adoption of investigated technologies is mainly related to companies declaring the following backshoring drivers: innovation, testing of technologies, and time-to-market reduction.

Of specific note is the Dachs et al.'s (2019) study, in which the authors point out that the higher level of responsiveness allowed by Industry 4.0 technologies may be carefully evaluated in terms of geographical distribution of firms' customers. More specifically, if company customers are located in countries/regions other than the home country, the adoption of Industry 4.0 technologies would induce companies to implement RTC strategies, either in the form of near-shoring or of further offshoring.

To sum up, the structured literature review conducted earlier offers a varied set of results which are not conclusive. While several authors recognize that single Industry 4.0 technologies (mainly 3D/additive manufacturing and automation) may have an impact on manufacturing relocation decisions, their impact is highly dependent on the strategic aims pursued by the company. Moreover, analyses have been focused, until now, on a restricted number of countries (mainly in Europe). Further investigations are then requested; in this respect, evidence belonging to the UMRD—which will be discussed in the next section—may contribute to the academic debate.

3.2 Empirical Findings

The literature review did not provide homogeneous results that can be considered conclusive; therefore, to further investigate the topic we now analyze empirical evidence from the UMRD. The latter includes data collected from secondary sources of backshoring decisions performed by European and US companies. Up to the end of December 2018, the UMRD covered a total of 1279 instances of evidence regarding backshoring decisions implemented by companies belonging to 24 European countries (814), the USA (428), and other foreign countries (37). Before analyzing the impact (if any) of Industry 4.0 technologies on the backshoring decisions, it seems useful to point out the main characteristics of the sampled backshoring decisions. In so doing, similarities and differences among the two main subsamples (European vs. US companies) deserve specific attention.

As far as the geographical dimension (home vs. host country/region) is concerned, three out of four US companies backshored from Asia (in particular from China), while European companies implemented backshoring more homogeneously from Asia and Europe (Table 2). Moreover, it is worth noting that the majority of intra-Europe relocations have been implemented among Western countries, i.e., among high-cost nations (when compared with those in Eastern Europe).

The breakdown by firm's size shows a higher homogeneity among the two subsamples, even if large companies are slightly more overrepresented in the European one (52.2% of total ones vs 43.7%).

Focusing the attention on industries, among most representative industries both in Europe and in the USA there is "manufacture of electrical equipment" and "manufacture of machinery and equipment not elsewhere classified n.e.c." (difference up to 1%). Differently, "manufacture of leather and related products" is an industry in which more European companies implemented the relocation, while "manufacture of computer, electronic, and optical products" is more diffused in the USA.

Examining drivers of relocation (Table 3), for both European and US companies three of the four most important drivers are related to the value-based quadrants of the Fratocchi et al.'s (2016) framework, namely: "customer responsiveness

Table 2 Breakdown of backshoring decisions by home and host region

Host region/country	Europe (%)	USA (%)	Others (%)	World (%)
China	33.8	61.0	45.9	43.2
Asia (other than China)	9.2	11.0	8.1	9.8
Asia (not specified)	2.2	4.0	5.4	2.9
Asia	*45.2*	*75.9*	*59.5*	*55.9*
Eastern Europe and former USSR	17.6	0.7	10.8	11.7
Western Europe	26.0	7.7	21.6	19.8
Europe (not specified)	0.5	0.2		0.4
Europe and the former USSR	*44.1*	*8.6*	*32.4*	*31.9*
North Africa and the Middle East	3.7	0.9		2.7
South Africa	0.1	0.0		0.1
Africa (not specified)	0.2	0.2		0.2
Africa	*4.1*	*1.2*		*3.0*
USA	0.4	0.0		0.2
North America (not USA)	2.0	2.3	2.7	2.1
Central and South America	1.5	9.8	2.7	4.3
Americas	*3.8*	*12.1*	*5.4*	*6.6*
Oceania	*0.1*	*0.2*		*0.2*
Not available	*2.7*	*1.9*	*2.7*	*2.4*
Total	*100.0*	*100.0*	*100.0*	*100.0*

Source: UnivAQ More reshoring dataset

Table 3 Breakdown of backshoring decisions by declared motivation[a]

Motivation	Europe (%)	USA (%)	Others (%)	World (%)
Customer responsiveness/vicinityHigher service quality	17.1	27.6	8.1	20.3
Logistics costs (including freight costs)	16.3	28.5	8.1	20.2
Made in effect (home country)	19.5	22.2	5.4	20.0
Delivery time (including delays)	17.3	24.5	10.8	19.5
Offshored poor product quality	16.1	23.1	13.5	18.4
Firm's organizational restructuring	17.6	8.6	16.2	14.5
Adoption of automation and/or other innovative product/process technologies (excluding 3D printing/additive manufacturing)	**13.1**	**11.7**	**13.5**	**12.7**
Increasing labor cost in the host country (including higher productivity in the home country)	7.6	19.4	18.9	11.9
Total cost of ownership	11.8	11.4	16.2	11.8

Source: UnivAQ More reshoring dataset
[a]Motivations declared by at least 10% of companies at the worldwide level. Motivation belonging to Industry 4.0 in bold

Table 4 Backshoring motivations cited jointly with the production automation[a]

Motivations	Europe (%)	USA (%)	Others (%)	World (%)
Customer responsiveness/vicinity Higher service quality	12.9	13.6	66.7	13.8
Logistic costs (including freight costs)	8.3	8.2	33.3	8.5
Made in effect (home country)	18.9	10.5		15.6
Delivery time (including delays)	18.4	8.6	50.0	14.8
Offshored poor product quality	13.7	10.1	40.0	12.8
Cost and difficulties in controlling the host country activities	21.8	19.4	50.0	21.4
Vicinity of engineering and production + Firm's strategies focused on product and process innovations	17.2	12.7		15.0

Source: UnivAQ More reshoring dataset
[a]Only motivations cited by at least 10% of companies at the worldwide level

improvement" (20.3%), "made in effect" (20% of total sample), and "delivery time" (19.5%) while "logistics costs" (20.2%) belongs to the cost quadrants. Even if these drivers are relevant for both the subsamples, they were slightly more cited by US companies.

When considering Industry 4.0 technologies, findings by Ancarani et al. (2019) are confirmed since companies only cited production automation and additive manufacturing as drivers for relocation to the home country. However, while automation has been declared a backshoring driver in 12.7% of the sampled decisions (with a slight over-citation by European companies: 13.1% vs. 11.7%), the adoption of additive manufacturing technologies has been considered as a reshoring motivation in only 1.3% of the sampled relocation decisions. Moreover, such a technology has been implemented almost exclusively by US companies (0.5% vs. 2.8%). Finally, only five (four US and one European) out of the 16 firms adopting 3D/additive manufacturing technologies also cited product automation as a driver for the backshoring decision. This finding confirms—at least partially—the Ancarani and Di Mauro (2018) and Ancarani et al.'s (2019) evidence that the two investigated technologies are likely to support different typologies of reshoring decisions. More specifically, both articles suggest that production automation is more consistent with "cost-oriented" and "flexibility-oriented" backshoring decisions while "quality-oriented" ones are better supported by additive manufacturing technologies. However, our data unexpectedly show that—considering only the 10 most cited motivations—production automation has been jointly cited with the following three motivations (all referring to quality-oriented relocation decisions): "cost and difficulties in controlling the host country activities" (21.4% of total companies cite this motivation), "made in effect" (15.6%), and "vicinity of engineering and production" (15%). In contrast, issues regarding production costs (e.g. "total cost of ownership" and "labor costs/productivity") are jointly cited by less than 10% of the sampled companies adopting production automation (Table 4).

Table 5 Backshoring evidence citing production automation: breakdown by firm's size

Firm's size	% of total European companies	% of total US companies	% of total other countries' companies	% of worldwide companies
Large	11.5	9.6	10.3	10.9
Medium	18.2	12.6		16.2
Small and micro	12.2	13.3	25.0	12.8
n.a.		33.3	100.0	16.7
Total	13.1	11.7	13.5	12.7

Source: UnivAQ More reshoring dataset

Table 6 Backshoring evidence citing production automation: breakdown by host region

Host region	% of European companies	% of US companies	% of other countries' companies	% of worldwide companies
Asia	13.6	12.3	13.6	13.0
Europe and the former USSR	12.8	13.5		12.5
Africa	21.2			18.4
Americas	3.2	9.6		7.1
Oceania				
Not available	13.6		100.0	12.9
Total	13.1	11.7	13.5	12.7

Source: UnivAQ More reshoring dataset

Though 3D/additive manufacturing technologies have been cited as a backreshoring driver by very few companies (16 out of 1,269), it is worth noting that companies citing such a technology mainly stated their backshoring decisions were based on "cost and difficulties in controlling the host country activities" (5.8% of total companies cited this motivation) and "vicinity of engineering and production" (4%). This finding is consistent with the expectations of Ancarani and Di Mauro, Fratocchi, Orzes, and Sartor (2018), and Ancarani et al. (2019).

Given the little evidence of backshoring decisions implementing 3D/additive manufacturing technologies, further insights may emerge when considering the breakdown of backshoring decisions citing product automation as a driver by size, geography, and industry. As far as size is concerned (Table 5), quite unexpectedly data show this technology—which generally requires high levels of investment—to be mainly adopted by medium-sized companies (16.2% of total firms in the range vs. 10% for the large ones and 12.8% for small and micro ones), especially among European companies.

When considering the geographic issues (Table 6), data clearly show that the adoption of automated production technologies is not influenced by the host region where companies have earlier offshored production activities. Also, this finding is partially unexpected, since one would have expected that backshoring decisions regarding production activities located in low-cost countries (e.g., Asia) would be

Table 7 Backshoring evidence citing production automation: breakdown by firm's industry[a]

NACE Code	Description	Number of companies at the worldwide level	% of total European companies	% of total US companies	% of total other countries' companies	%
26	Manufacture of computer, electronic, and optical products	153	15.9	7.8	14.3	12.4
28	Manufacture of machinery and equipment n.e.c.	130	11.6	9.8		10.8
27	Manufacture of electrical equipment	128	6.4	17.8		10.2
14	Manufacture of apparel	108	16.3	14.8		15.7
25	Manufacture of fabricated metal products, except machinery and equipment	85	26.2	17.1		21.2
22	Manufacture of rubber and plastic products	73	13.5	12.1	100.0	16.4
10	Manufacture of food products	58	22.4			19.0
31	Manufacture of furniture	52	22.2	4.0		13.5
24	Manufacture of basic metals	21	31.3			23.8

Source: UnivAQ More reshoring dataset
[a]Only industries with no less than 20 companies at the worldwide level

largely supported by automation when compared with medium- and high-cost countries (e.g., Europe). Moreover, it is in contrast with previous findings of Dachs et al. (2019) in terms of higher "Industry 4.0 readiness" of large companies with respect to small and medium ones. A possible explanation for this unexpected result may be represented by latter-day implementation of automated production systems by the medium companies.

Finally, when considering the firms' industry (Table 7) dissimilarities among European and US backshoring decisions clearly emerge. For instance while only 7% of European leather manufacturers declared to have invested in production automation when backshoring, the corresponding value for US companies is 28.6%. In contrast, European companies have highly automated furniture production (22.2%) compared with US ones (4%). This finding seems to confirm that the home country—at least partially—matters when investigating the backshoring decisions (Wan, Orzes, Sartor, & Nassimbeni, 2019).

4 Concluding Remarks

The chapter aimed to investigate the relationship (if any) between Industry 4.0 technologies and decisions to relocate earlier offshored manufacturing activities to the home country. To shed new light on this research question, an exploratory approach has been implemented adopting a two-step methodology. First of all a structured literature review has been conducted on a sample of 115 Scopus indexed journal articles published between 2007 and August 2019. This research clearly shows the topic is attracting growing interest among scholars (at least from 2014). However, they mainly focus on specific technologies, namely production automation and 3D printing/additive manufacturing. In any case, findings are not sufficient to be conclusive and seem to be influenced by geographic issues, since automation is not equally implemented in the different Western countries, also because of their different industry structure (i.e., the type of sectors in which local companies operate). Only four journal articles specifically address the relationship between Industry 4.0 technologies and backshoring decisions; moreover, their findings are somewhat contrasting. For instance, Dachs et al. (2019) found a significant and positive relationship (but not also the causality) between the two issues while Ancarani et al. (2019) and Stentoft and Rajkumar (2019) did not discover any connection. This finding might induce the speculation that country-specific issues may influence the obtained results, since Dachs et al. (2019) focus on German, Austrian, and Swiss companies, while Stentoft and Rajkumar (2019) on Danish ones. As clearly showed by analyzing data from the UMRD, the European and US companies that backshored their production based on Industry 4.0 technologies are characterized by some dissimilarities, especially in terms of industry and adopted technology (production automation vs. additive manufacturing). Finally, the geographic dimension deserves a specific note since investments in Industry 4.0 technologies may be influenced by financial aids provided by national and/or local government bodies. In this respect, Ancarani et al. (2019) suggest policymakers should not only offer companies the possibility to reduce the fixed cost belonging to the adoption of Industry 4.0 technologies but also to develop "the necessary digital competencies for the successful exploitation of these technologies" (2018, p. 10). This is consistent with Nujen, Halse, et al. (2018) who state Industry 4.0 investments "have little value unless complemented with employee upskilling programs" (2018, see p. 690). Moreover, authors point out that the use of advanced technologies, as the ones belonging to Industry 4.0, needs to be complemented with other manufacturing competences. In this respect, Lampón and González-Benito (2019) state that backshoring strategies are more likely implemented by companies which improved their key manufacturing resources (e.g., process optimization, technologies, and facilities). Moreover, in the case of backshoring decisions coupled with re-insourcing ones, these competences may be already available within the firm or, more often, have to be redeveloped activating adequate learning process. To sum up, the effective implementation of both Industry 4.0 technologies and backshoring strategies requests companies to carefully evaluate their readiness and activate proper learning processes.

Another issue emerging as relevant is the one concerning the size. While it is generally expected Industry 4.0 technologies are more easily adopted by large companies, analysis of UMRD data provides evidence that—at least production automation—is mainly implemented by European medium-sized companies and US small and micro ones. Future research should further address this aspect, given the implications in terms of availability of skilled employees (Stentoft & Rajkumar, 2019).

A third question is still open as regards the relationships (if any) between the adoption of a specific Industry 4.0 technology and the strategic aims pursued by the backshoring decision. While Ancarani et al. (2019) and Ancarani and Di Mauro (2018) suggest that production automation is more consistent with "cost-oriented" and "flexibility-oriented" backshoring decisions; data from the UMRD provide evidence that companies adopting this technology were driven by motivations belonging to the "quality-oriented" backshoring decisions.

The previous discussion induces us to conclude that further studies are requested to further investigate the proposed research question. Our study has an explorative aim and is mainly based on secondary data; therefore, our conclusions are not generalizable. However, it may represent a useful state of the art of the academic debate and of backshoring evidence available up to now. In this respect, we suggest future research should couple a longitudinal case study approach with quantitative surveys.

Appendix

Publication year	Authors	Journal	Automated production system	Additive manufacturing	I4.0 and Back-shoring
2007	Kinkel, S., Lay, G., Maloca, S.	International Journal of Entrepreneurship and Small business			
2009	Kinkel, S., Maloca, S.	Journal of Purchasing and Supply Management			
2011	Hogg, D.	Manufacturing Engineering			
2012	Kinkel, S.	International Journal of Operations and Production Management			

(continued)

Year	Authors	Journal			
2013	Baldwin, R., Venables, A.J.	Journal of International Economics			
2013	Canham, S., Hamilton, R.T.	Strategic Outsourcing			
2013	Denning, S.	Strategy and Leadership			
2013	Ellram, L.M.	Journal of Supply Chain Management			
2013	Ellram, L.M., Tate, W.L., Petersen, K.J.	Journal of Supply Chain Management			
2013	Gray, J.V., Skowronski, K., Esenduran, G., Rungtusanatham, M.	Journal of Supply Chain Management			
2014	Arlbjørn, J.S., Mikkelsen, O.S.	Journal of Purchasing and Supply Management	X		
2014a	Bailey, D., De Propris, L.	Cambridge Journal of Regions, Economy and Society	X		
2014b	Bailey, D., De Propris, L.	Revue d'Economie Industrielle			
2014	Fratocchi, L., Di Mauro, C., Barbieri, P., Nassimbeni, G., Zanoni, A.	Journal of Purchasing and Supply Management			
2014	Kinkel, S.	Journal of Purchasing and Supply Management			
2014	Martínez-Mora, C., Merino, F	Journal of Purchasing and Supply Management			
2014	Mugurusi, G., de Boer, L.	Strategic Outsourcing			
2014	Tate, W.L.	Journal of Purchasing and Supply Management	X		
2014	Tate, W.L., Ellram, L.M., Schoenherr, T., Petersen, K.J.	Business Horizons	X		

(continued)

Year	Authors	Journal				
2014	Wu, X., Zhang, F.	Management Science				
2014	Zhai, W.	Economic Modelling				
2015	Ancarani, A., Di Mauro, C., Fratocchi, L., Orzes, G., Sartor, M.	International Journal of Production Economics	X			
2015	Belussi, F.	Investigaciones Regionales				
2015	Fox, S.	Technology on Society				
2015	Grandinetti, R., Tabacco, R.	International Journal of Globalisation and Small Business				
2015	Grappi, S., Romani, S., Bagozzi, R.P.	Journal of the Academy of Marketing Science				
2015	Gylling, M., Heikkilä, J., Jussilä, K., Saarinen, M.	International Journal of Production Economics				
2015	Razvadovskaja, YV., Shevcenko, I.K.	Asian Social Science				
2015	Sardar, S., Lee, Y.H.	Mathematical Problems in Engineering				
2016	Abbasi, H.	Journal of Textile and Apparel Technology and Management	X			
2016	Ashby, A.	Operations Management Research				
2016	Bals, L., Kirchoff, J.F., Foerstk, K.	Operations Management Research	X			
2016	Barbieri, P., Stentoft, J.	Operations Management Research	X			
2016	Foerstl, K., Kirchoff, Bals, L.	International Journal of Physical Distribution and Logistics Management	X			

(continued)

2016	Foster, K.	Journal of Textile and Apparel Technology and Management	X			
2016	Fratocchi, L., Ancarani, A., Barbieri, P., Di Mauro, C., Nassimbeni, G., Sartor, M., Vignoli, M., Zanoni, A.	International Journal of Physical Distribution and Logistics Management	X			
2016	Huq, F., Pawar, K. S., Rogers, H.	Production Planning and Control				
2016	Joubioux, C., Vanpoucke, E.	Operations Management Research				
2016	Lavissière, A., Mandjá, K., Fedi, L.	Supply Chain Forum				
2016	Młody, M.	Entrepreneurial Business and Economics Review				
2016	Moradlou, H., Backhouse, C.J.	Proceedings of the Institution of Mechanical Engineers, Part B: Journal of Engineering Manufacture				
2016	Presley, A., Meade, L., Sarkis, J.	Supply Chain Forum				
2016	Robinson, P.K., Hsieh, L.	Operations Management Research				
2016	Saki, Z.	Journal of Textile and Apparel Technology and Management	X			
2016	Sardar, S., Lee, Y. H., Memon, M.S.	Sustainability				
2016	Srai, J.S., Ané, C.	International Journal of Production Research	X			
2016a	Stentoft, J., Mikkelsen, O.S., Jensen, J.K.	Operations Management Research	X			

(continued)

2016b	Stentoft, J., Mikkelsen, O.S., Jensen, J.K.	Supply Chain Forum	X			
2016a	Stentoft, J., Ohlager, J., Heikkilä, J., Thoms, L.	Operations Management Research	X			
2016	Sutherland et al.	CIRP Annals - Manufacturing Technologies				
2016	Uluskan, M., Joines, J.A., Godfrey, A.B.	Supply Chain Management				
2016	Zhai, W., Sun., S, Zhang, G.	Operations Management Research	X			
2017	Benstead, A. V., Stevenson, M., Hendry, L.C.	Operations Management Research	X			
2017	Bettiol, M., Burlina, C., Chiarvesio, M., Di Maria, E.	Investigaciones Regionales				
2017	Brandon-Jones, E., Dutordoir, M., Frota Neto, J.Q., Squire, B.	Journal of Operations Management				
2017	Bye, E., Erickson, K.	Research Journal of Textile and Apparel				
2017	Chen, L., Hu, B.	Manufacturing and Service Operations Management				
2017	Delis, A., Driffield, N., Temouri, Y.	Journal of Business Research				
2017	Fel, F., Griette, E.	Strategic Direction				
2017	Gray, J.V., Skowronski, K., Esenduran, G., Rungtusanatham, M. et al.	Journal of Operations Management				
2017	Hartman, P.L., Ogden, J.A., Withlin, J.R., Hazen, B.T.	Business Horizons				
2017	Moradlou, H., Backhouse, C. J., Ranganathan, R.	International Journal of Physical Distribution and Logistics Management	X			

(continued)

Year	Authors	Journal				
2017	Schmidt, A.S.T., Touray, E., Hansen, Z. N. L.	Production Engineering				
2017	Tate, W.L., Bals, L.	International Journal of Physical Distribution and Logistics Management	X			
2017	Uluskan, M., Godfrey, A. B., & Joines, J. A.	Journal of the Textile Institute	X			
2017	Wiesmann, B., Snoei, J.R., Hilletofth, P., Eriksson, D.	European Business Review	X			
2017	Yegul et al.	Computers and industrial engineering				
2017	Zhao, L., Huchzermeier, A.	European Journal of Operational Research	X			
2018	Ancarani, A., Di Mauro, C.	IEEE Engineering Management Review	X	X		X
2018	Bailey, D., Corradini, C., De Propris, L.	Cambridge Journal of Economics				
2018	Baraldi, E., Ciabuschi, F., Lindahl, O., Fratocchi, L.	Industrial Marketing Management				
2018	Barbieri, P., Ciabuschi, F., Fratocchi, L., Vignoli, M.	Journal of Global Operations and Strategic Sourcing	X	X		
2018	Boffelli, A., Golini, R., Orzes, G., Dotti, S.	IEEE Engineering Management Review				
2018	Di Mauro, C., Fratocchi, L., Orzes, G., Sartor, M.	Journal of Purchasing and Supply Management	X			
2018	Engström, G., Hilletofth, P., Eriksson, D., Sollander, K.	World Review of Intermodal transportation research	X			
2018	Engström, G., Sollander, K., Hilletofth, P., Eriksson, D.	Journal of Global Operations and Strategic Sourcing	X			

(continued)

Year	Authors	Journal				
2018a	Fratocchi, L.	World Review of Intermodal Transportation Research		X		
2018	Grappi, S., Romani, S., Bagozzi, R.P.	Journal of World Business				
2018	Hasan, R.	Journal of Textile and Apparel Technology and Management	X	X		
2018	Heikkilä, J., Nenonen, S., Olhager, J., Stentoft, J	World Review of Intermodal Transportation Research	X			
2018	Heikkilä, J., Martinsuo, M., Nenonen, S.	Journal of Manufacturing Technology Management	X			
2018a	Johansson, M., Olhager, J.	International Journal of Production Economics				
2018b	Johansson, M., Olhager, J.	Journal of Manufacturing Technology Management				
2018	Moore, M.E., Rothenberg, L., Moser, H.	Journal of Manufacturing Technology Management	X	X		
2018	Moradlou, H., Tate, W.	World Review of Intermodal Transportation Research		X		
2018	Nujen, B.B., Halse, L.L., Damm, R., Gammelsæter, H.	Journal of Manufacturing Technology Management	X	X		
2018	Pal, Harper, Vellesalu	The International Journal of Logistic Management	X			
2018	Sirilertsuwan, P., Ekwall, D., Hjelmgren, D.	International Journal of Logistics Management				
2018	Stentoft, J., Mikkelsen, O. S., Jensen, J. K., Rajkumar, C.	International Journal of Production Economics	X			

(continued)

2018	Theyel, G., Hofman, K., Gregory, M.	Economic Development Quarterly				
2018	Vanchan, V., Mulhall, R., Bryson, J.	Growth and Change	X	X		
2018	Yu, U.-J., Kim, J.-H.	Journal of Fashion Marketing and Management				
2018	Nujen, B.B., Mwesiumo, D.E., Solli-Sæther, H., Slyngstad, A.B., Halse, L.L.	Journal of Global Operations and Strategic Sourcing	X	X		
2019	Ancarani, A., Di Mauro, C., Mascali, F.	Journal of World Business	X	X	X	
2019	Barbieri, P., Elia, S., Fratocchi, L., Golini, R.	Journal of Purchasing and Supply Management	X			
2019	Ciabuschi, F., Lindahl, O., Barbieri, P., Fratocchi, L.	European Business Review				
2019	Dachs, B., Kinkel, S., Jäger, A., Palčič, I.	Journal of Purchasing and Supply Management				
2019	Fjellstrom, D., Fang, T., Chimenson, D.	Journal of Asia Business Studies	X			
2019	Gadde, L.E., Jonsson, P.	Journal of Purchasing and Supply Management				
2019	Hilletofth, P, Eriksson, D., Tate, W., Kinkel, S.	Journal of Purchasing and Supply Management				
2019	Hilletofth, P., Sequeira, M., Adlemo, A.	Expert Systems with Applications				
2019	Oshri, I., Sidhu, J. S., Kotlarsky, J.	Journal of Business Research				
2019	Johansson, M., Olhager, J., Heikkilä, J., Stentoft, J.	Journal of Purchasing and Supply Management				
2019	Luthra, S., Mangla, S.K., Yadav, G.	Journal of Cleaner Production				

(continued)

Year	Authors	Journal			
2019	Perrone, G., Bruccoleri, M., Mazzola, E.	International Journal of Production Economics			
2019	Mohiuddin, M., Rashid, M.D.M., Al Azad, M.D.S., Su, Z.	International Journal of Logistics Research and Applications			
2019	Orzes, G., & Sarkis, J.	Resources, Conservation & Recycling			
2019	Piatanesi, B., Arauzo-Carod, J. M.	Growth and Change			
2019	Sayem, A., Feldman, A., Ortega-Mier, M.	BRQ Business Research Quarterly	X		
2018	Talamo, G., Sabatino, M.	Contemporary Economics			
2019	Thakur-Werns, P.	Journal of Global Operations and Strategic Sourcing			
2019	Wan, L. Orzes, G., Sartor, M., Di Mauro, C., Nassimbeni, G.	Journal of Purchasing and Supply Management			
2019	Wan, L. Orzes, G., Sartor, M., Nassimbeni, G.	Journal of Purchasing and Supply Management			
2019	Dachs, B., Kinkel, S., Jäger, A.	Journal of World Business	X		X
2019	Stentoft, J., Rajkumar, C.	International Journal of Production Research	X		X
Total			42	10	4

References

Abbasi, M. H. (2016). It's not offshoring or reshoring but right-shoring that matters. *Journal of Textile and Apparel, Technology and Management, 10*(2), 1–6.

Alcácer, J., Cantwell, J., & Piscitello, L. (2016). Internationalization in the information age: A new era for places, firms, and international business networks? *Journal of International Business Studies, 47*, 499–512.

Ancarani, A., & Di Mauro, C. (2018). Reshoring and Industry 4.0: How often do they go together? *IEEE Engineering Management Review, 46*(2), 87–96.

Ancarani, A., Di Mauro, C., Fratocchi, L., Orzesc, G., & Sartorc, M. (2015). Prior to reshoring: A duration analysis of foreign manufacturing ventures. *International Journal of Production Economics, 169*, 141–155.

Ancarani, A., Di Mauro, C., & Mascali, F. (2019). Backshoring strategy and the adoption of Industry 4.0: Evidence from Europe. *Journal of World Business, 54*(4), 360–371.

Arlbjørn, J. S., & Mikkelsen, O. S. (2014). Backshoring manufacturing: Notes on an important but under-researched theme. *Journal of Purchasing & Supply Management, 20*(1), 60–62.

Ashby, A. (2016). From global to local: Reshoring for sustainability. *Operations Management Research, 9*(3-4), 1–14.

Bailey, D., Corradini, C., & De Propris, L. (2018). Home-sourcing and closer value chains in mature economies: The case of Spanish manufacturing. *Cambridge Journal of Economics, 42*(6), 1567–1584.

Bailey, D., & De Propris, L. (2014a). Manufacturing reshoring and its limits: The UK automotive case. *Cambridge Journal of Regions, Economy and Society, 20*(1), 66–68.

Bailey, D., & De Propris, L. (2014b). Reshoring: Opportunities and limits for manufacturing in the UK – the case of the auto sector. *Revue D'économie Industrielle, 1*(145), 45–61.

Baldwin, R., & Venables, A. J. (2013). Spiders and snakes: Offshoring and agglomeration in the global economy. *Journal of International Economics, 90*(2), 245–254.

Bals, L., Kirchoff, J. F., & Foerstl, K. (2016). Exploring the reshoring and insourcing decision making process: Toward an agenda for future research. *Operations Management Research, 9*(3-4), 1–15.

Baraldi, E., Ciabuschi, F., Lindahl, O., & Fratocchi, L. (2018). A network perspective on the reshoring process: The relevance of the home- and the host-country contexts. *Industrial Marketing Management, 70*, 156–166.

Barbieri, P., Ciabuschi, F., Fratocchi, L., & Vignoli, M. (2018). What do we know about manufacturing reshoring? *Journal of Global Operations and Strategic Sourcing, 11*(1), 79–122.

Barbieri, P., Elia, S., Fratocchi, L., & Golini, R. (2019). Relocation of second degree: Moving towards a new place or returning home? *Journal of Purchasing and Supply Management.* https://doi.org/10.1016/j.pursup.2018.12.003

Barbieri, P., & Stentoft, J. (2016). Reshoring: A supply chain innovation perspective. *Operations Management Research, 9*(3-4), 49–144.

Belussi, F. (2015). The international resilience of Italian industrial districts/clusters (ID/C) between knowledge re-shoring and manufacturing off (near-)shoring. *Investigaciones Regionales – Journal of Regional Research, 32*, 89–113.

Benstead, A. V., Stevenson, M., & Hendry, L. C. (2017). Why and how do firms reshore? A contingency-based conceptual framework. *Operations Management Research, 10*(3–4), 85–103.

Bettiol, M., Burlina, C., Chiarvesio, M., & Di Maria, E. (2017). From delocalisation to backshoring? Evidence from Italian industrial districts. *Investigaciones Regionales, 39*, 137–154.

Bettiol, M., Burlina, C., Chiarvesio, M., & Di Maria, E. (2018). Manufacturing, where art thou? Value chain organization and cluster-firm strategies between local and global. In V. De Marchi, E. De Maria, & G. Gereffi (Eds.), *Local clusters in global value chains* (pp. 155–174). London: Routledge.

Boffelli, A., Golini, R., Orzes, G., & Dotti, S. (2018). "How to reshore": Some evidence from the apparel industry. *IEEE Engineering Management Review, 46*, 4. https://doi.org/10.1109/EMR.2018.2886183

Brandon-Jones, E., Dutordoir, M., Frota Neto, J. Q., & Squire, B. (2017). The impact of reshoring decisions on shareholder wealth. *Journal of Operations Management, 49-51*, 31–36.

Brennan, L., Ferdows, K., Godsell, J., & Golini, R. (2015). Manufacturing in the world: Where next? *International Journal of Operations & Production Management, 35*(9), 1253–1274.

Brynjolfsson, E., & McAfee, A. (2014). *The second machine age: Work, progress, and prosperity in a time of brilliant technologies.* New York: Norton.

Buckley, P. J. (2011). International integration and coordination in the global factory. *Management International Review, 51*, 269–283.

Buckley, P. J., & Ghauri, P. N. (2004). Globalisation, economic geography and the strategy of multinational enterprises. *Journal of International Business Studies, 35*(2), 81–98.

Bye, E., & Erickson, K. (2017). Opportunities and challenges for Minnesota sewn product manufacturers. *Research Journal of Textile and Apparel, 21*(1), 72–83.

Canham, S., & Hamilton, R. T. (2013). SME internationalisation: offshoring, 'backshoring', or staying at home in New Zealand. *Strategic Outsourcing: An International Journal, 6*(3), 277–291.

Chen, L., & Hu, B. (2017). Is reshoring better than offshoring? The effect of offshore supply dependence. *Manufacturing & Service Operations Management, 19*(2), 166–184.

Chen, W., & Kamal, F. (2016). The impact of information and communication technology adoption on multinational firm boundary decisions. *Journal of International Business Studies, 47*(5), 563–576.

Ciabuschi, F., Lindahl, O., Barbieri, P., & Fratocchi, L. (2019). Manufacturing reshoring: A strategy to manage risk and commitment in the logic of the internationalization process model. *European Business Review, 31*(1), 139–159.

d'Aveni, R. (2015). The 3-D printing revolution. *Harvard Business Review, 93*(5), 40–48.

Dachs, B., Kinkel, S., & Jäger, A. (2019). Bringing it all back home? Backshoring of manufacturing activities and the adoption of Industry 4.0 technologies. *Journal of World Business, 54*(6), 101017.

De Backer, K., DeStefano, T., & Menon, C., J.R Suh (2018). *Industrial robotics and the global organisation of production* (Working Papers 2018/03). Paris: OECD Science, Technology and Industry.

Delis, A., Driffield, N., & Temouri, Y. (2017). The global recession and the shift to re-shoring: Myth or reality? *Journal of Business Research, 103*, 632–643.

Denning, S. (2013). Boeing's offshoring woes: Seven lessons every CEO must learn. *Strategy & Leadership, 41*(3), 29–35.

Di Mauro, C., Fratocchi, L., Orzes, G., & Sartor, M. (2018). Offshoring and backshoring: A multiple case study analysis. *Journal of Purchasing and Supply Management, 24*(2), 108–134.

Ellram, L. M. (2013). Offshoring, reshoring and the manufacturing location decision. *Journal of Supply Chain Management, 49*(2), 3–6.

Ellram, L. M., Tate, W. L., & Petersen, K. J. (2013). Offshoring and reshoring: An update on the manufacturing location decision. *Journal of Supply Chain Management, 49*(2), 14–22.

Engström, G., Hilletofth, P., Eriksson, D., & Hilletofth, P. (2018). Drivers and barriers of reshoring in the Swedish manufacturing industry. *World Review of Intermodal Transportation Research, 7*(3), 195–220.

Engström, G., Sollander, K., Hilletofth, P., & Eriksson, D. (2018). Reshoring drivers and barriers in the Swedish manufacturing industry. *Journal of Global Operations and Strategic Sourcing, 11*(2), 174–201.

Fel, F., & Griette, E. (2017). Near-reshoring your supplies from China: A good deal for financial motives too. *Strategic Direction, 33*(2), 24–26.

Fjellstrom, D., Fang, T., & Chimenson, D. (2019). Explaining reshoring in the context of Asian competitiveness: Evidence from a Swedish firm. *Journal of Asia Business Studies, 13*(2), 277–293.

Foerstl, K., Kirchoff, J. F., & Bals, L. (2016). Reshoring and insourcing: drivers and future research directions. *International Journal of Physical Distribution & Logistics Management, 46*(5), 492–515.

Foster, K. (2016). A prediction of US knit apparel demand: Making the case for reshoring manufacturing investment in new technology. *Journal of Textile and Apparel, Technology and Management, 10*(2), 1–10.

Fox, S. (2015). Moveable factories: How to enable sustainable widespread manufacturing by local people in regions without manufacturing skills and infrastructure. *Technology in Society, 42*, 49–60.

Fratocchi, L. (2018a). Additive manufacturing technologies as a reshoring enabler: A why, where and how approach. *World Review of Intermodal Transportation Research, 7*(3), 264–293.

Fratocchi, L. (2018b). *Additive manufacturing as a reshoring enabler considerations on the why issue*. In Paper presented at workshop on metrology for Industry 4.0 and IoT, MetroInd 4.0 and IoT 2018 – Proceedings 6 August 2018, Article number 8428316.

Fratocchi, L., Ancarani, A., Barbieri, P., Di Mauro, C., Nassimbeni, G., Sartor, M., et al. (2015). Manufacturing back-reshoring as a nonlinear internationalization process. In R. Van Tulder, A. Verbeke, & R. Drogendijk (Eds.), *The future of global organizing, Progress in international business research (PIBR)* (pp. 367–405). Bingley: Emerald.

Fratocchi, L., Ancarani, A., Barbieri, P., Di Mauro, C., Nassimbeni, G., Sartor, M., et al. (2016). Motivations of manufacturing back-reshoring: An interpretative framework. *International Journal of Physical Distribution & Logistics Management, 46*(2), 98–127.

Fratocchi, L., Di Mauro, C., Barbieri, P., Nassimbenid, G., & Zanoni, A. (2014). When manufacturing moves back: Concepts and questions. *Journal of Purchasing and Supply Management, 20*(1), 54–59.

Gadde, L. E., & Jonsson, P. (2019). Future changes in sourcing patterns: 2025 Outlook for the Swedish textile industry. *Journal of Purchasing and Supply Management, 25*(3), 100526.

Grandinetti, R., & Tabacco, R. (2015). A return to spatial proximity: Combining global suppliers with local subcontractors. *International Journal of Globalisation and Small Business, 7*(2), 139–161.

Grappi, S., Romani, S., & Bagozzi, R. P. (2015). Consumer stakeholder responses to reshoring strategies. *Journal of the Academy of Marketing Science, 43*(4), 453–471.

Grappi, S., Romani, S., & Bagozzi, R. P. (2018). Reshoring from a demand-side perspective: Consumer reshoring sentiment and its market effects. *Journal of World Business, 53*(2), 194–208.

Gray, J. V., Esenduran, G., Rungtusanatham, M. J., & Skowronski, K. (2017). Why in the world did they reshore? Examining small to medium-sized manufacturer decisions. *Journal of Operations Management, 49-51*, 37–51.

Gray, J. V., Skowronski, K., Esenduran, G., & Skowronski, K. (2013). The reshoring phenomenon: What supply chain academics ought to know and should do. *Journal of Supply Chain Management, 49*(2), 27–33.

Greenwood, M. (2011). *Which business and management journal database is best?* Accessed December 22, 2018, from https://bizlib247.wordpress.com/2011/06/19/which-business-and-management-journal-database-is-best/

Gylling, M., Heikkilä, J., Jussila, K., & Saarinen, M. (2015). Making decisions on offshore outsourcing and backshoring: A case study in the bicycle industry. *International Journal of Production Economics, 162*, 92–100.

Hartman, P. L., Ogden, J. A., Wirthlin, J. R., & Hazen, B. T. (2017). Nearshoring, reshoring, and insourcing: Moving beyond the total cost of ownership conversation. *Business Horizons, 60*(3), 363–373.

Hasan, R. (2018). Reshoring of US apparel manufacturing: Lesson from an Innovative North Carolina based manufacturing company. *Journal of Textile and Apparel, Technology and Management, 10*(4), 1–6.

Heikkilä, J., Martinsuo, M., & Nenonen, S. (2018). Backshoring of production in the context of a small and open Nordic economy. *Journal of Manufacturing Technology Management, 29*(4), 658–675.

Heikkilä, J., Nenonen, S., Olhager, J., & Nenonen, S. (2018). Manufacturing relocation abroad and back: Empirical evidence from the Nordic countries. *World Review of Intermodal Transportation Research, 7*(3). https://doi.org/10.1504/WRITR.2018.10014279

Hilletofth, P., Eriksson, D., Tate, W., & Kinkel, S. (2019). Right-shoring: Making resilient offshoring and reshoring decisions. *Journal of Purchasing and Supply Management, 25*(3), 100540. https://doi.org/10.1016/j.pursup.2019.100540

Hilletofth, P., Sequeira, M., & Adlemo, A. (2019). Three novel fuzzy logic concepts applied to reshoring decision-making. *Expert Systems with Applications, 126*, 133–143.

Hogg, D. (2011). Lean in a changed world. *Manufacturing Engineering, 147*(3), 102–113.

Huq, F., Pawar, K. S., & Rogers, H. (2016). Supply chain configuration conundrum: How does the pharmaceutical industry mitigate disturbance factors? *Production Planning & Control, 27*(14), 1206–1220.

Johanson, M., & Olhager, J. (2018a). Comparing offshoring and backshoring: The role of manufacturing site location factors and their impact on post-relocation performance. *International Journal of Production Economics, 205*, 37–46.

Johanson, M., & Olhager, J. (2018b). Manufacturing relocation through offshoring and backshoring: The case of Sweden. *Journal of Manufacturing Technology Management, 29*(4), 637–657.

Johansson, M., Olhager, J., Heikkilä, J., & Stentoft, J. (2019). Offshoring versus backshoring: Empirically derived bundles of relocation drivers, and their relationship with benefits. *Journal of Purchasing and Supply Management, 25*(3), 100509.

Joubioux, C., & Vanpoucke, E. (2016). Towards right-shoring: A framework for off- and re-shoring decision making. *Operations Management Research, 9*(3-4), 1–16.

Kagermann, H., Wahlster, W., & Helbig, J. (2013). *Recommendations for implementing the strategic initiative INDUSTRIE 4.0: Securing the future of German manufacturing industry..* Final report of the Industrie 4.0 Working Group. Berlin: Acatech – Deutsche Akademie der Technikwissenschaften e.V.

Ketokivi, M., Turkulainen, V., Seppälä, T., Rouvinend, P., & Ali-Yrkköd, J. (2017). Why locate manufacturing in a high-cost country? A case study of 35 production location decisions. *Journal of Operations Management, 49*, 20–30.

Kinkel, S. (2012). Trends in production relocation and backshoring activities: Changing patterns in the course of the global economic crisis. *International Journal of Operations & Production Management, 32*(6), 696–720.

Kinkel, S. (2014). Future and impact of backshoring – some conclusions from 15 years of research on German practices. *Journal of Purchasing and Supply Management, 20*(1), 63–65.

Kinkel, S., Lay, G., & Maloca, S. (2007). Development, motives and employment effects of manufacturing offshoring of German SMEs. *International Journal of Entrepreneurship and Small Business, 4*(3), 256–276.

Kinkel, S., & Maloca, S. (2009). Drivers and antecedents of manufacturing offshoring and backshoring – a German perspective. *Journal of Purchasing and Supply Management, 15*(3), 154–165.

Lampón, J. F., & González-Benito, J. (2019). Backshoring and improved manufacturing resources in firms' home location. *International Journal of Production Research.* https://doi.org/10.1080/00207543.2019.1676479

Laplume, A. O., Petersen, B., & Pearce, J. M. (2016). Global value chains from a 3D printing perspective. *Journal of International Business Studies, 47*, 595–609.

Lavissière, A., Mandják, T., & Fedi, L. (2016). The key role of infrastructure in backshoring operations: The case of free zones. *Supply Chain Forum: An International Journal, 17*(3), 143–155.

Leamer, E. E., & Storpe, M. (2001). The economic geography of the Internet age. *Journal of International Business Studies, 32*(4), 641–665.

Lu, Y. (2017). Industry 4.0: "A survey on technologies, applications and open research issues". *Journal of Industrial Information Integration, 6*, 1–10.

Luthra, S., Mangla, S. K., & Yadav, G. (2019). An analysis of causal relationships among challenges impeding redistributed manufacturing in emerging economies. *Journal of Cleaner Production, 225*, 949–962.

Manning, S. (2014). Mitigate, tolerate or relocate? Offshoring challenges, strategic imperatives and resource constraints. *Journal of World Business, 49*(4), 522–535.

Martínez-Mora, C., & Merino, F. (2014). Offshoring in the Spanish footwear industry: A return journey? *Journal of Purchasing and Supply Management, 20*(4), 225–237.

Młody, M. (2016). Backshoring in light of the concepts of divestment and de-internationalization: Similarities and differences. *Entrepreneurial Business and Economics Review, 4*(3), 167–180.

Mohiuddin, M., Rashid, M. M., Al-Azad, S. M., & Su, Z. (2019). Back-shoring or re-shoring: Determinants of manufacturing offshoring from emerging to least developing countries (LDCs). *International Journal of Logistics Research and Applications, 22*(1), 78–97.

Moore, M. E., Rothenberg, L., & Moser, H. (2018). Reshoring manufacturing in the textile and apparel industry. *Journal of Manufacturing Technology Management, 29*(6), 1025–1041.

Moradlou, H., & Backhouse, C. J. (2016). A review of manufacturing re-shoring in the context of customer-focused postponement strategies. *Proceedings of the Institution of Mechanical Engineers, Part B: Journal of Engineering Manufacture, 230*(9), 1561–1571.

Moradlou, H., Backhouse, C. J., & Ranganathan, R. (2017). Responsiveness, the primary reason behind re-shoring manufacturing activities to the UK: An Indian industry perspective. *International Journal of Physical Distribution & Logistics Management, 47*(2-3), 222–236.

Moradlou, H., & Tate, W. (2018). Reshoring and additive manufacturing. *World Review of Intermodal Transportation Research, 7*(3), 241–263.

Mugurusi, G., & De Boer, L. (2014). Conceptualising the production offshoring organisation using the viable systems model (VSM). *Strategic Outsourcing: An International Journal, 7*(3), 275–298.

Müller, J., Dotzauer, V., Voigt, K. I. (2017). Industry 4.0 and its impact on reshoring decisions of German manufacturing enterprises. In Bode C, Bogaschewsky R, Eßig, M R Lasch, W Stolzle (Eds.) *Supply chain research*. Heidelberg: Springer.

Nujen, B. B., Halse, L. L., Damm, R., & Gammelsaeter, H. (2018). Managing reversed (global) outsourcing – the role of knowledge, technology and time. *Journal of Manufacturing Technology Management, 29*(4), 676–698.

Nujen, B. B., Mwesiumo, D. E., Solli-Sæther, H., & Slyngstad, A. B. (2018). Backshoring readiness. *Journal of Global Operations and Strategic Sourcing, 12*(1), 172–195.

Orzes, G., & Sarkis, J. (2019). Reshoring and environmental sustainability: An unexplored relationship? *Resources, Conservation and Recycling, 141*, 481–482.

Oshri, I., Sidhu, J. S., & Kotlarsky, J. (2019). East, west, would home really be best? On dissatisfaction with offshore-outsourcing and firms' inclination to backsource. *Journal of Business Research, 103*, 644–653.

Pal, R., Harper, S., & Vellesalu, A. (2018). Competitive manufacturing for reshoring textile and clothing supply chains to high-cost environment: A Delphy study. *The International Journal of Logistics Management, 11*(1), 79–122.

Perrone, G., Bruccoleri, M., & Mazzola, E. (2019). The curvilinear effect of manufacturing outsourcing and captive-offshoring on firms' innovation: The role of temporal endurance. *International Journal of Production Economics, 211*, 197–210.

Piatanesi, B., & Arauzo-Carod, J. M. (2019). Backshoring and nearshoring: An overview. *Growth and Change, 50*, 806–823.

Pisano, G. P., & Shih, W. C. (2012). *Producing prosperity: Why America needs a manufacturing renaissance*. Boston: Harvard Business Press.

Presley, A., Meade, L., & Sarkis, J. (2016). A strategic sourcing evaluation methodology for reshoring decisions. *Supply Chain Forum: An International Journal, 17*(3), 156–169.

Razvadovskaja, Y. V., & Shevcenko, I. K. (2015). Dynamics of metallurgic production in emerging countries. *Asian Social Science, 11*(19), 178–184.

Robinson, P. K., & Hsieh, L. (2016). Reshoring: A strategic renewal of luxury clothing supply chains. *Operations Management Research, 9*(3-5), 1–13.

Saki, Z. (2016). Disruptive innovations in manufacturing–an alternative for re-shoring strategy. *Journal of Textile and Apparel, Technology and Management, 10*(2), 1–7.

Sardar, S., & Lee, Y. H. (2015). Analysis of product complexity considering disruption cost in fast fashion supply chain. *Mathematical Problems in Engineering, 2015*, 670831.

Sardar, S., Lee, Y., & Memon, M. (2016). A sustainable outsourcing strategy regarding cost, capacity flexibility, and risk in a textile supply chain. *Sustainability, 8*(3), 234.

Sayem, A., Feldmann, A., & Ortega-Mier, M. (2019). Investigating the influence of network-manufacturing capabilities to the phenomenon of reshoring: An insight from three case studies. *BRQ Business Research Quarterly, 22*(1), 68–82.

Schmidt, A. S. T., Touray, E., & Hansen, Z. N. L. (2017). A framework for international location decisions for manufacturing firms. *Production Engineering, 11*(6), 703–713.

Seuring, S., & Gold, S. (2012). Conducting content-analysis based literature reviews in supply chain management. *Supply Chain Management: An International Journal, 17*(5), 544–555.

Sirilertsuwan, P., Ekwall, D., & Hjelmgren, D. (2018). Proximity manufacturing for enhancing clothing supply chain sustainability. *The International Journal of Logistics Management, 29*(4), 1346–1378.

Srai, J. S., & Ané, C. (2016). Institutional and strategic operations perspectives on manufacturing reshoring. *International Journal of Production Research, 54*(23), 1–19.

Stentoft, J., Jensen, K. W., Philipsen, K., & Haug. A. (2019). Drivers and barriers for Industry 4.0 readiness and practice: A SME perspective with empirical evidence. In: *Proceedings of the 52nd Hawaii International Conference on system Sciences*, Hawaii (pp. 5155–5164).

Stentoft, J., Mikkelsen, O. S., & Jensen, J. K. (2016a). Flexicurity and relocation of manufacturing. *Operations Management Research, 9*(3/4), 1–12.

Stentoft, J., Mikkelsen, O. S., & Jensen, J. K. (2016b). Offshoring and backshoring manufacturing from a supply chain innovation perspective. *Supply Chain Forum: An International Journal, 17*(4), 190–204.

Stentoft, J., Mikkelsen, O. S., Jensen, J. K., & Rajkumar, C. (2018). Performance outcomes of offshoring, backshoring and staying at home manufacturing. *International Journal of Production Economics, 199*, 199–208.

Stentoft, J., Olhager, J., Heikkilä, J., & Thoms, L. (2016). Manufacturing backshoring: A systematic literature review. *Operations Management Research, 9*(3-4), 53–61.

Stentoft, J., & Rajkumar, C. (2018). Balancing theoretical and practical relevance in supply chain management research. *International Journal of Physical Distribution & Logistics Management, 48*(5), 504–523.

Stentoft, J., & Rajkumar, C., (2019). The relevance of Industry 4.0 and its relationship with moving manufacturing out, back and staying at home. *International Journal of Production Research*. https://doi.org/10.1080/00207543.2019.1660823

Strange, R., & Zucchella, A. (2018). Industry 4.0, global value chains and international business. *Multinational Business Review, 25*(3), 174–184.

Sutherland, J. W., Richter, J. S., Hutchins, M. J., Dornfeld, D., Dzombak, R., Mangold, J., et al. (2016). The role of manufacturing in affecting the social dimension of sustainability. *CIRP Annals, 65*(2), 689–712.

Talamo, G., & Sabatino, M. (2018). Reshoring in Italy: A recent analysis. *Contemporary Economics, 12*(4), 381–398.

Tate, W. L. (2014). Offshoring and reshoring: US insights and research challenges. *Journal of Purchasing and Supply Management, 20* (1), 66–68.

Tate, W. L., & Bals, L. (2017). Outsourcing/offshoring insights: Going beyond reshoring to rightshoring. *International Journal of Physical Distribution & Logistics Management, 47* (2-3), 106–113.

Tate, W. L., Ellram, L. M., Schoenherr, T., & Petersen, K. J. (2014). Global competitive conditions driving the manufacturing location decision. *Business Horizons, 57*(3), 381–390.

Thakur-Wernz, P. (2019). A typology of backsourcing: Short-run total costs and internal capabilities for re-internalization. *Journal of Global Operations and Strategic Sourcing, 12*(1), 42–61.

Theyel, G., Hofmann, K., & Gregory, M. (2018). Understanding manufacturing location decision making: Rationales for retaining, offshoring, reshoring and hybrid approaches. *Economic Development Quarterly, 32*(4), 300–312.

Uluskan, M., Godfrey, A. B., & Joines, J. A. (2017). Impact of competitive strategy and cost-focus on global supplier switching (reshore and relocation) decisions. *Journal of the Textile Institute, 108*(8), 1308–1318.

Uluskan, M., Joines, J. A., & Godfrey, A. B. (2016). Comprehensive insight into supplier quality and the impact of quality strategies of suppliers on outsourcing decisions. *Supply Chain Management: An International Journal, 21*(1), 92–102.

Vanchan, V., Mulhall, R., & Bryson, J. (2018). Repatriation or reshoring of manufacturing to the US and UK: Dynamics and global production networks or from here to there and back again. *Growth and Change, 49*(1), 97–121.

Wan, L., Orzes, G., Sartor, M., Di Mauro, C., & Nassimbeni, G. (2019). Entry modes in reshoring strategies: An empirical analysis. *Journal of Purchasing and Supply Management, 25*(3), 100522.

Wan, L., Orzes, G., Sartor, M., & Nassimbeni, G. (2019). Reshoring: Does home country matter? *Journal of Purchasing and Supply Management, 25*(4), 100551.

Wiesmann, B., Snoei, J. R., Hilletofth, P., & Eriksson, D. (2017). Drivers and barriers to reshoring: A literature review on offshoring in reverse. *European Business Review, 29*(1), 15–42.

Wu, X., & Zhang, F. (2014). Home or overseas? An analysis of sourcing strategies under competition. *Management Science, 60*(5), 1223–1240.

Yegul, M. F., Erenay, F. S., Striepe, S., & Yavuz, M. (2017). Improving configuration of complex production lines via simulation-based optimization. *Computers & Industrial Engineering, 109*, 295–312.

Yu, U. J., & Kim, J. H. (2018). Financial productivity issues of offshore and "Made-in-USA" through reshoring. *Journal of Fashion Marketing and Management, 22*(3), 317–334.

Zhai, W. (2014). Competing back for foreign direct investment. *Economic Modelling, 39*, 146–150.

Zhai, W., Sun, S., & Zhang, G. (2016). Reshoring of American manufacturing companies from China. *Operations Management Research, 9*(3–4), 1–13.

Zhao, L., & Huchzermeier, A. (2017). Integrated operational and financial hedging with capacity reshoring. *European Journal of Operational Research, 260*(2), 557–570.

Knowledge and Digital Strategies in Manufacturing Firms: The Experience of Top Performers

Marco Bettiol, Mauro Capestro, Eleonora Di Maria, and Stefano Micelli

Abstract In the past few decades, ICT supported firms managing knowledge through both codification and social interaction, also at distance. Within the Industry 4.0 framework, firms can access knowledge through the cloud and rely on big data and AI to improve their processes and enhance their market comprehension. However, it is not fully explored how knowledge management should be organized in the fourth industrial revolution, since a lot of emphasis has been given to automatization in data management, while the relational dimension of knowledge management has received limited attention. Through an empirical analysis based on mixed method of a survey on 75 top performing Italian manufacturing firms and follow-up on 5 case studies, the chapter explores these questions to identify the implications of Industry 4.0 for firms' strategy.

1 Introduction

Knowledge is a strategic component of the modern firm (Drucker, 1995; Kogut & Zander, 1992; Nonaka & Takeuchi, 1995; Spender, 1996), and it is at the core of the elements that distinguishes one firm from another. Knowledge is idiosyncratic and firm-specific. The way firms manage knowledge is not just a matter of efficiency but it is crucial to compete in the markets and to sustain its competitive advantage. Alavi and Leidner (2001) affirm that *"Because knowledge-based resources are usually difficult to imitate and socially complex, the knowledge-based view of the firm posits that these knowledge assets may produce long-term sustainable competitive advantage"* (p. 107). This approach is rooted in the resource-based view of the firm that

M. Bettiol · M. Capestro · E. Di Maria (✉)
Department of Economics and Management 'Marco Fanno', University of Padova, Padova, Italy
e-mail: eleonora.dimaria@unipd.it

S. Micelli
Department of Management, Ca' Foscari University, Venice, Italy

emphasizes the importance of how tangible resources are combined and used by virtue of the firm's know-how (Barney, 1991).

For those reasons, knowledge management (KM) literature plays an important role in the firm's strategy. The origin of the strategic importance of knowledge management can be traced back to Polanyi *"I shall reconsider human knowledge by starting from the fact that we can know more than we can tell."* (1966, p. 4). This interpretation of knowledge as largely based on a tacit and unarticulated dimension at both individual and organizational levels has led to the definition of managerial practices for transforming the knowledge in a way that could be used by the firm. As Nonaka and Takeuchi maintain in their seminal book *The Knowledge Creating Company* (1995), the firm could improve its competitiveness leveraging on the distributed knowledge pools within its boundaries. The well-established SECI model elaborated by Nonaka and Takeuchi is a way to get access to *individual-tacit* knowledge and transform that in explicit and useful knowledge for the firm. The interplay between tacit and explicit knowledge is at the heart of the strategic relevance of knowledge.

Several KM initiatives conceived technology, in particular, information technology (IT), with the purpose of *"extracting"* knowledge and make this contextual and highly personal resource at the disposal of the firm as a whole. Around the 2000s, in coincidence with the rapid growth of the Internet and the development of new software for storing and managing information, KM focused on the use of those applications as *filters* through which information produced within the firm is captured and stored. KM projects implemented database and software for storing and processing information (i.e., Enterprise Resource Planning) with the idea that knowledge can be interpreted as oil,[1] an extraordinarily valuable resource distributed within the firm that just need to be discovered and put in the pipeline to be ready to use. Ironically, the same oil metaphor will come back later when we discuss the potential of Industry 4.0 in relation to big data.

Although those KM initiatives were effective for data and the management of simple tasks, they had difficulties in fostering knowledge in the firm (Davenport & Prusak, 1998). Database can store a lot of data and information, but it is questionable how those data and information can be transformed into knowledge especially when the process within the firm is complex. Indeed, some authors underline that tacit and explicit knowledge are strictly interconnected and that it is hard to separate one from the other (Alavi & Leidner, 2001). Tacit and explicit are two sides of the same coin. Brown and Duguid affirm (2000): *"From the idea that tacit knowledge is "non-tradable" and needs to be converted into explicit form to circulate, we come instead to the idea not only that conversion (if it involves uprooting knowledge from the tacit) is problematic, but also that tacit knowledge is required to make explicit knowledge usefully tradable or mobile. Only by first spreading the practice in relation to which the explicit makes sense is the circulation of explicit knowledge*

---

[1]https://www.economist.com/leaders/2017/05/06/the-worlds-most-valuable-resource-is-no-longer-oil-but-data

worthwhile (Cook and Brown, 1999). **Knowledge, in short, runs on rails laid by practice."** *(p. 204, bold is ours)*. It seems a paradox that to transform tacit into explicit more tacit knowledge is required.

Knowledge is not a treasure that waits to be discovered, but it is a more subtle object that needs a social context to be produced and shared. Paraphrasing a well-known book by Brown and Duguid (2000, 2001), we could say that information and knowledge have a social life: it relies, in other words, in a social fabric and a common understanding. Knowledge is not a regular good that can be transferred and produced mechanically, but it requires a social context and shared sense-making mechanisms.

From this perspective, the role of technology changes: from knowledge extraction to an enabler of collaboration and sharing among workers within and outside the firm. ICT (Information and Communications Technologies) supports humans in the production, memorization, sharing, and application of knowledge via tools that enhance collaboration and foster networking. The social and cultural facets of knowledge are not discarded and become central to the development of technologies. In particular, communication technologies take center stage. Conversions among people are crucial for sharing information and problem-solving. As confirmed by ethnographic research, humans produce and exchange knowledge through narratives and interactions (Orr, 1996). In this regard, a technology family called Groupware, composed of forum, discussion bulletin, email, etc. was widely used in KM projects for sustaining the interaction among workers. The objective was to foster the development of communities of workers within the firm to increase knowledge circulation.

The comparison between an extractive approach to KM practices based on IT and relational approach to KM practices based on communication technologies is useful to consider the new technological frontier of Industry 4.0 and its promise to have both an increased amount of data available from the production of an item to its consumption and new software capabilities (artificial intelligence) for processing information. The potential of this new technology (AI) is to transform traditional manufacturing and to create innovative services for the customer. Thanks to machine learning and deep learning, software can create knowledge in an automated way that could lead to better decision-making. From this perspective, Industry 4.0 is not only a new family of technologies for supporting knowledge production and sharing among people but also—and here is the novelty—an independent source of knowledge although generated algorithmically. At least potentially, machines could have the ability to put information into practice taking decision with limited human supervision (Floridi, 2016). Those technologies have the capability of acting in the physical world, facing and solving problems as robots that auto determine malfunctions and suggest possible interventions. More information and more computational power available seem to lead to a knowledge revolution that will change the way we produce and share knowledge and also introducing new agents in the knowledge management field, machines with increasing information process capabilities.

If, as we mentioned at the beginning of this paragraph, knowledge is a strategic resource, firms have to deal with this revolution and to use the new potentiality offered by technology to sustain their competitive advantage. Although there is great emphasis on Industry 4.0 in the media and in the consulting world, it is unclear how and when firms will adopt those new technologies. More importantly, it is still questionable with what KM perspective those technologies will be used by firms. Are the firms investing in Industry 4.0 in order to automatize and extract knowledge or they prefer to increase the communication and relations among workers? Are firms focusing on more tacit or explicit forms of knowledge? How autonomous are those machines and how they are changing decision-making?

In order to answer those questions, we conducted quantitative and qualitative research on Italian manufacturing firms. We decided to focus on manufacturing because this industry is at the cusp of a great transformation led by those new technologies. We selected Italy because it is the second-largest manufacturing country in Europe and it is mainly based on low/medium-tech productions that expose Italian firms to the aggressive competitiveness not only by low-cost countries, but also from more developed ones that are becoming more flexible and innovative in their production with the help of such new technologies. To understand how Italian firms are dealing with Industry 4.0, we decided to focus on the best performing firms that we thought to have the higher probability of using those technologies compared to other firms.

Before analyzing the result of our research, it is useful to take a deeper look at what are Industry 4.0 technologies and how they promise to transform manufacturing and KM within the firm.

2 Manufacturing and Industry 4.0

2.1 *Managing Knowledge to Support Manufacturing*

It is difficult to draw a line and define when this revolution took place. One good starting point is "How to (Make) Almost Anything" the title of a famous engineering class taught by Prof. Neil Gershenfeld at MIT. The class was specifically designed for applying the potential of digital technologies to the physical world. As Neil Gershenfeld (2012) affirms: "A new digital revolution is coming, this time in fabrication. It draws on the same insights that led to the earlier digitizations of communication and computation, but now what is being programmed is the physical world rather than the virtual one. Digital fabrication will allow individuals to design and produce tangible objects on demand, wherever and whenever they need them. Widespread access to these technologies will challenge traditional models of business, foreign aid, and education." (p. 43) CNC machines, 3D printers, and the distribution of cheap sensors are the protagonist of a remarkable transformation of a physical object into digital information or bits and back from bits to atoms. "Atoms become the new bits," as Chris Anderson (2010) put it, and it is possible to shape objects following the rules of the digital while overcoming the limitations of traditional manufacturing. For example, 3D printing can produce shapes that are impossible to obtain with the traditional techniques of subtracting manufacturing (milling machines).

But that revolution is more profound. Gershenfeld admonishes that the very nature of this technological transformation is based on quantity and quality of the information available: "The revolution is not additive versus subtractive manufacturing; it is the ability to turn data into things and things into data" (p. 44). More information means more precision, efficiency, and a decreasing cost of manufacturing and, at the same time, more flexibility thanks to the use of this information for producing an increased variety of products. Indeed, this revolution aims at solving one of the most important trade-offs in traditional manufacturing: volume vs. personalization, or between quantity vs. quality. With existing technology, mass production needs product standardization and economy of scale, while customization is economically possible in low volume and at the expense of a high cost of production. Ideally, that revolution could lead to a future of makers (Anderson, 2010) that can self-produce customized products—based on their needs—at a fraction of the cost of today. From this perspective, the manufacturing industry is no more necessary in increasing the autonomy of the users in (auto) making the products. The focus of attention moves from industry to individuals that can now own the means of production (i.e., 3D printing) and access the required knowledge for making products via online communities of users (Anderson, 2012). We could define that as the American approach to that technological revolution.

In Europe, that technological revolution took the name of Industry 4.0 a term coined in Germany (Kagermann, Helbig, Hellinger, & Wahlster, 2013; Lasi et al., 2014) as part of a public initiative for understanding the impacts of automation in manufacturing. As the largest manufacturing country in Europe, Germany was interested in maintaining its leadership in the industry applying the potential of digital technologies. Instead of conceiving a future without manufacturing production, Germans worked on the idea of transforming the manufacturing process thanks to the new possibility offered by digital technologies and its integration with traditional machines. The starting point of the German approach is based on the concept of cyber-physical systems that aims at managing the interconnections between physical assets and computational capabilities (Lee, Bagheri, & Kao, 2015). The new availability of cheap digital sensors that can be distributed in the manufacturing process and the possibility of connecting isolated machines to a computer network increase the quantity and quality of data and information available for the firm (Wang, Törngren, & Onori, 2015). Thanks to the extensive use of connected machines and the increasing amount of data, the factory itself can become smarter, able—*at least theoretically*—to self-organize production based on continuous feedback (Wang, Wan, Zhang, Li, & Zhang, 2016).

Although there are several possible definitions of Industry 4.0, the literature on engineering and manufacturing tried to identify the main technologies that compose Industry 4.0. Based on an extensive literature review, Alcacer and Cruz-Machado (2019) consider the following technologies under the umbrella of Industry 4.0: *the Industrial Internet of Things* (sensors and connected machines), *Cloud Computing* (distributed platform for accessing information and computation), *Big Data* (storage

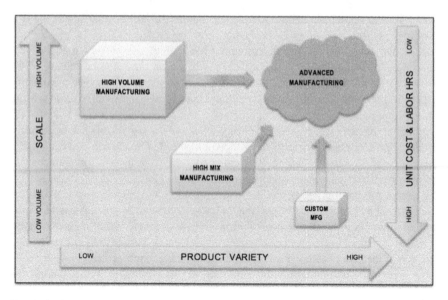

Fig. 1 Shifting trade-offs with advanced manufacturing: scale, product variety, and unit costs. Source: Sturgeon, Fredriksson, and Korka (2017)

of increasing amount of data on manufacturing processes), *Simulation* (the possibility to anticipate the results of a manufacturing process through specialized software), *Augmented Reality* (the possibility to help the operator on the line via enriched information mixed with the perception of the real situation), *Additive Manufacturing* (i.e., 3D printers), *Horizontal and Vertical Integration Systems* (integration of different technologies within different areas of the firm—horizontal—or in the supply chain—vertical), *Autonomous Robots* (robots that can take decisions and can cooperate among them with limited human supervision), and *Cybersecurity* (mechanism to protect data and computation systems from external aggressions). To those technologies we could also add the increased availability of *artificial intelligence* (AI) solutions that based on the data gathered could define a new course of action and taking decisions with limited human control (Fry, 2018). The list of technologies has also been identified within specific policies developed in many countries (i.e., Germany, Italy) to financially support firms' technological investments in selected directions.

The objective of Industry 4.0 is to respond to the increasing request for personalized products coming from the consumers. In this context, smart manufacturing is synonymous for flexible and agile manufacturing that can respond more quickly and precisely to the market. The technological revolution is opening a new scenario of advanced manufacturing where custom products are feasible at decreasing cost (see Fig. 1). The anticipated and never achieved so far *mass customization* (Pine II, 1993) seems finally at hand. The technology and its application at the factory level seem mature enough to make mass customization real. From the perspective of Industry 4.0, the factory is the epicenter of the revolution where the potential of technology can be fully deployed. The factory is where the strategic knowledge is produced and stored.

Although they have differences, both the American and the European (German) take on technological revolution have something in common: the increasing demand of knowledge that is needed in manufacturing. We could say that beyond being a technological revolution it is also a cognitive one, in the sense that new knowledge and understanding are needed. There are at least three main areas where we expect that *knowledge* will expand.

The first one is *knowledge* of the *product*. Sensors, digital machines, computer networks, databases, software, etc. are producing an increasing amount of data and information available about how and when a product is used by the consumer. In this perspective, the rise of smart products (IoT) can open a new domain for a better understanding of the needs of the consumers. Porter and Heppelmann (2014) sustain that the diffusion of smart products could also modify the structure of value chains and the rules of competition. The consumer is not just the end user of the product but could be the new starting point of the production and could be involved in the definition of the product through online collaboration or co-produce the product herself (Anderson, 2012). The knowledge developed by the customer in her own experience could be useful for defining new business models centered on the consumer (Bogers, Hadar, & Bilberg, 2016).

The knowledge of the product means also a better understanding of the production processes. Although it is questionable that more data can be translated into knowledge per se, the possibility to gather information about the product itself and the machines used in the factory could help the operator to increase their knowledge and to have new sources for problem-solving and improving the production process. For example, the possibility of adding a sensor to a traditional milling machine could give to the operator and the plant manager a better understanding of the defects of production and this may lead to new maintenance practices or a new organization of production. This is even more true if we consider the complexity of existing value chains where the production is fragmented among several firms specialized in the specific phase of production. This is particularly relevant in the context of small and medium-sized enterprises (SMEs) and of clusters where a high division of labor exists among localized firms and innovation is tightly coupled with supplier–buyer interaction in the value chain (Chiarvesio, Di Maria, & Micelli, 2004).

The second one is *knowledge* of the *technology*. Although several of the technologies that are part of the Industry 4.0 are not new, their combination is something that is not well established (Alcacer & Cruz-Machado, 2019). The power of digital technologies (sensors, database, software) combined with more flexible machines (robots, additive manufacturing) is a new paradigm that is literally in the making. Specific knowledge on the single technological domain is probably abundant, but how those technologies interact and could be something new that has to be perfected. Best practices on how to mix and match those technologies are still under development and it will be a learning-by-doing experience. At the moment there is still a lot of confusion on when, how, and where to apply those technologies. The recent failure of Adidas in developing their project of a highly automatized and digitalized factory called *Speedfactory* is indicative of technological systems that are not mature. In particular, according to what has emerged in the media, the robotic factory was able just to produce a limited number of models, which mainly consisted of running shoes with a knit upper while it was unable to produce leather ones. "It's a different kind of joining process behind it where we just don't have a solution yet,"[2] said Ulrich Steindorf, senior director of manufacturing at Adidas. Just because a specific technological solution is available, it does not mean that it could be applied to a specific production process. There is a lot of dark spots to explore within the technological framework of Industry 4.0. Dead ends and best practices are not well known yet.

The third one is *knowledge* of the *management* of the firm. We refer to the combination between the new technological features and new opportunities to be discovered. As happened in the previous industrial revolutions, the introduction of new technologies implied a new way of organizing both the production and the definition of the product. It took several years after the invention and diffusion of electricity before an entrepreneur such as Henry Ford developed an organizational model based on the assembly line and large scale of production in order to take full advantage of that technological innovation. Besides, Ford had to identify a new market opportunity: a car that was targeted to the mass instead of small niches of affluent consumers. That concept was something completely new for the time.

Technology needs to meet strategy to express its full potential. If the analogy with the second industrial revolution holds, human creativity in the form of firms' strategy is still important to implement Industry 4.0. Several authors affirm that firms adopt the new technologies because they expect to achieve some specific results in the areas of manufacturing as of marketing to improve their competitiveness (Bharadwaj, El Sawy, Pavlou, & Venkatraman, 2013; Kane, Palmer, Phillips, Kiron, & Buckley, 2015). From this perspective, the adoption requires new knowledge that depends on the business purposes firms aim to achieve. Recent research shows that firms chose to adopt Industry 4.0 technologies for specific strategic motivations such as the improvement of efficiency (Bonfanti, Del Giudice, & Papa, 2018) and productivity (Yao & Lin, 2016) and the reorganization of manufacturing activities, with the opportunity to have them locally (Müller, Kiel, & Voigt, 2018). Other main drivers of adoption relate to the achieving of market and marketing benefits (Coreynen, Matthyssens, & Van Bockhaven, 2017), such as the improvement of customer service as the product variety (Leeflang, Verhoef, Dahlström, & Freundt, 2014).

---

[2]The digital magazine *Quarz* reported the words of Ulreich Steindorf https://qz.com/1746152/adidas-is-shutting-down-its-speedfactories-in-germany-and-the-us/

2.2 From Data to AI to Enhanced Learning

In May 2017, *The Economist* dedicated its cover to the big tech giants such as Apple, Google, Facebook, Microsoft Tesla, and Uber and their dominance of the digital market. The title of the cover story was symbolically: "The world's most valuable resource is no longer oil, but data".[3] Being the article focused on the monopolist power of such tech giants, data were considered at the base of the success of those companies. In other words, data are the new scarce resource; the one who has access to it will dominate the market. Since 2017, this metaphor becomes very popular and was used by several opinion leaders, like Jaron Lanier,[4] to highlight the potential negative consequences of the use of data by tech giants.

Besides those critical notes on privacy and monopoly, "data as oil" was extensively used in the large consultancy firms to advise companies to invest in data management to take advantage of the potential of AI. Indeed, the leap forward that AI did last year is remarkable. In the field of voice recognition, image recognition, and language translation, the progress of AI is tangible and has led to the deployment of very powerful service that is at our fingertips. For example, you can easily translate a text written in a foreign language by simply pointing the camera of the smartphone over the text. Even more significant are the performances of IBM Watson and Alpha Go of Google that outperformed their human counterpart at games like Chess or Go and at the TV quiz *Jeopardy*. Those results are the outputs of a powerful combination of new algorithm techniques based on deep learning and the availability of big data, giant accumulation of data produced by users in the digital platform like social media, credit card transactions, medical information, etc..

If data is the oil, AI is the modern refinery able to distill knowledge out of the raw but valuable material. Several technology vendors marketed the (almost) unlimited potential of the application of AI for solving big human problems like cancer or the development of new solutions for fighting climate change. That trust on the potential benefit of the application of AI entered also in the consultancy world that is sponsoring the development of AI initiatives among manufacturing and service firms. Although several authors (Tegmark, 2017; Zuboff, 2019) warned about the potential threats that the extensive use of the AI could have for our society, it is beyond the scope of this chapter to analyze those possible negative consequences in detail.

The debate of the potential of AI in elaborating data and producing knowledge is not new and goes back to the 1980s and 1990s at the time of the application of the so-called expert systems (Davenport, 2019). Technology is now more powerful and there is greater availability of data but, as Davenport (2019) admonishes, the question remains the same: how to make AI solutions at work at the firm level.

---

[3]The original article could be reached at this link https://www.economist.com/leaders/2017/05/06/the-worlds-most-valuable-resource-is-no-longer-oil-but-data

[4]The Privacy Project was developed by Jerome Lanier for the *New York Times* https://www.nytimes.com/interactive/2019/09/23/opinion/data-privacy-jaron-lanier.html

Most of the solutions now available are conceived not for tailored applications in the firm but for more general purposes. As Davenport reported, many projects of AI are facing hard times when they are used in real business processes.

The literature on KM (Pauleen, 2017; Pauleen & Wang, 2017) warned that gathering bigger data does not necessarily lead to more knowledge because knowledge is the outcome of sense-making and human judgment. As we saw in the Introduction, the old problem of tacit and explicit knowledge seems to come back when we try to apply AI and big data into practice. Probably, the oil metaphor is misleading. The fact the data are relatively abundant, although not distributed evenly, does not turn necessarily into better solutions as several negative case studies demonstrated. In her book *Hello World: Being Human in the Age of Algorithms*, Hannah Fry (2018) reported many problematic cases in the use of AI such as Steve Talley's, an ordinary American citizen mistaken by FBI facial recognition software for a dangerous bank robber. Steve Talley was brutally arrested, suffered several injuries (some serious), and spent 2 months in a maximum-security prison and it took more than a year to be rehabilitated. Another example are AI applications that are used daily in American courts to decide the amount of punishment based on the probability of recidivism. It is always the popular jury that decides but hardly contradicts the algorithm's response. The result is that black defendants are more likely to remain in jail because they are considered at greater risk of recidivism. The problem here is related to the data on which the algorithm is based, which is biased by the fact that historically in the USA blacks are more arrested than whites. That disproportion in the starting data is reflected in a higher probability in the calculation of the recidivism potential.

As the philosopher Luciano Floridi pointed out (2016), we have too much trust in the intelligence of AI and, on the contrary, we should think that AI is rather a divorce between intelligence and agency. Floridi's take on this is that AI machines dramatically increased their capability of an action in the real world, but this is happening without much contextual intelligence. They can do some tasks, but those tasks need to be very well defined although the system is not able to adjust to the variations that the real context of use can have. In other words, complexity needs to be reduced to let AI thrive and this is not always possible.

Instead of considering AI as a substitution for human intelligence, we should consider AI as an important tool for sustaining learning at the level of individuals and organizations. From this perspective, AI can complement human intelligence and can give us different points of view on events and on decisions to take. They can multiply alternatives and help us to take better decisions. Humans and algorithms can live together, helping each other. When this is happening, the results are remarkable. As Fry (2018) reports, one of the most convincing example of mutual learning and collaboration is on the judgment of cancer cells. AI helps pathologists by reducing the number of suspicious areas to be examined and leaving the final decision to doctors. As Fry says, "The algorithm never gets tired and the pathologist is rarely wrong. The man-machine collaboration in this case leads to an incredible level of accuracy of 99.5%!"

Taking into account this complex and yet to be defined scenario of KM, we aim at exploring how Industry 4.0 technologies are shaping KM in manufacturing firms, the motivations of adoption, impacts in terms of product and process innovation, and knowledge creation within the competitive framework of the firm.

3 Empirical Analysis: Methodology and Results

3.1 Methodology

To reach the research purposes, the study focuses on medium and large Italian companies named *Champions* according to the economic and financial performance criteria of selection defined by *ItalyPost*—Italian Study Centre (Zovico, 2018). In particular, from a population of 14,632 companies between 20 and 120 million euros in turnover 500 were identified that meet, in addition to the turnover range, the following requirements: (1) CAGR (compound annual growth rate) 2010–2016 higher than 7%; (2) EBITDA average of the last 3 years greater than or equal to 10%; (3) debt ratio lower than or equal to 80%; (4) net debt/EBITDA average of the last 3 years lower than or equal to 80%; (5) number of employees greater than 20; and (6) a positive net income 2016. In this way the analysis focuses on a sample of medium and large firms usually engaged in the knowledge creation and management processes for the success of business (McAadam & Reird, 2001) with high performance that may assure no financial constraints that may negatively affect the adoption of new technologies (Kamble, Gunasekaran, & Dhone, 2019).

3.2 Measures

For the research objectives, we adopted a mixed method with quantitative and qualitative analyses. For the quantitative analysis, we carried out a survey submitting a structured questionnaire through computer-assisted web interview (CAWI) methodology (appropriate for contacting a large sample) to entrepreneurs, chief operations officers, or managers in charge of manufacturing and technological processes. The survey was carried out in the period October 2018–March 2019. The questionnaire aimed at assessing some of the enabling technologies that shape the fourth industrial revolution (Tortorella and Fettermann, 2018), specifically (1) autonomous robots, (2) additive manufacturing, (3) big data, (4) cloud, (5) artificial intelligence, (6) augmented reality, and (7) IoT and intelligent products. In addition, we also evaluated the use of some digital technologies typically by artisans in Italy for the deployment of a 3D digital model (Bonfanti et al., 2018), such as laser cutting and 3D scanner. Through a yes-no dichotomous measure, we asked respondents if firms have adopted or not each one of the selected technologies investigated. The choice of these types of technologies is in line with the Italian Ministry of Economic

Development regulation that, in 2016, delimited the scope of Industry 4.0 to the new technologies enabling the advanced manufacturing systems and the cyber-physical system (see Agostini & Filippini, 2019).

In addition to firm descriptive characteristics and the evaluation of Industry 4.0 and ICT endowment, the questionnaire assessed other strategic variables such as the motivations of adoption and the impact of new technologies on business results, on product performance, and on working skills and methods. According to recent literature (Ancarani, Di Mauro, & Mascali, 2019; Dalenogare et al., 2018; Müller et al., 2018; Schneider, 2018; Stentoft & Rajkumar, 2019; Whysall, Owtram, & Brittain, 2019), we considered the most common drivers of adoption as well as the benefits of the new digital technologies to evaluate the variable before mentioned. Specifically the motivations of adoption as well as the impact in terms of business results refer to (1) efficiency and productivity, (2) product diversification and customization, (3) new marketing opportunities, (4) international competitiveness, (5) reshoring and backshoring of production activities, (6) customer service, (7) respond to market requests (customer and standard industry), and (8) the aspect of environmental sustainability. The motivations of adoption were measured with a 5-point Likert scale (from 1 = *not at all* to 5 = *very much*). Instead, the impact of Industry 4.0 technologies on business was measured through a dichotomous variable (*yes-no*).

The impact on terms of product use and development refers to (1) the development of product-related services, (2) the role of customer in design and production processes, and (3) the control over product use and the distribution process. Finally, with respect to the working changes related to Industry 4.0, we assessed the modifications in terms of working methods and specifically about the relationships among the different business areas (production and others principally) and with suppliers as well as the creation of new knowledge for both product and production improvements. Impacts on product and on working changes were assessed through a 5-point Likert scale ranging from 1 (*not at all*) to 5 (*very much*).

To analyze the importance of all variables with respect to the Industry 4.0 strategy and explore the relationship with knowledge management, we transformed the Likert variables in dichotomous variables coding with 1 the highest values of Likert scale, which are 4 and 5, and considering 0 all the other three values, which are 1, 2, and 3.

3.3 Sample Descriptive

Through the survey on the 500 *Champions* companies, we were able to collect 75 questionnaires (15% of the population). Table 1 reports the description of the sample. Firstly, *Champions* are international companies characterized by a high export rate (60.5%), but with production activities and suppliers rooted locally (same company region and/or Italy). They focus on customized/customizable

Table 1 Descriptive statistics

Turnover 2017 (average; euros)	59.1 million
Employees 2017 (average)	
Total	138.4
Production	75.6
R&D	9.5
Marketing	3.7
Export 2017 (average; % on turnover)	60.5% (*first export country: 26.7%*)
R&D expenditure 2017 (average; % on turnover)	6.4%
Market	
Business-to-business	64%
Business-to-consumer	36%
Production output	
Standard products	42.8%
Bespoke products	27.5%
Customized products	29.7%
Production activities location	
Same company region	62.7%
Italy	22.4%
Abroad	14.9%
Suppliers' location	
Same company region	31.8%
Italy	45.3%
Abroad	22.9%
Competitive factors	
Product quality	31.1%
Product innovation	27.9%
Production flexibility	16.4%
Customer service	11.5%
Production efficiency	4.9%
Design	1.6%

$N = 75$

products for the larger part of production output, aiming mainly for product quality and innovation and production flexibility as competitive factors.

As far as the technology endowment is concerned, both referred to the ICT as to the Industry 4.0. Figs. 2 and 3 present interesting results. Firstly, Fig. 2 highlights that companies show a good ICT endowment especially as regards the technologies for the business and processes management (like ERP, CAD/CAM). The web environment is restricted to website and social media as not all firms use e-commerce as a selling platform. Moreover, Fig. 2 shows that *Champions* are technological companies as the 79% of the sample already use at least four (median) ICTs.

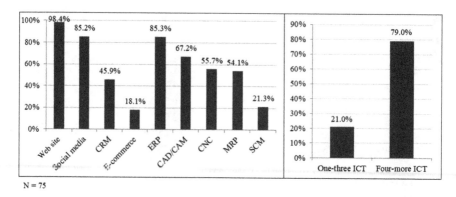

Fig. 2 ICT endowment

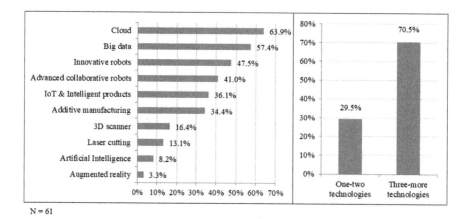

Fig. 3 Industry 4.0 adoption

Secondly, as far as the Industry 4.0 technologies are concerned, 81.3% (61 on 75) of the sample adopted at least one of the new technologies investigated. Figure 3 shows that *Champions* adopted two main sets of technologies: the technologies enabling the data management (Cloud, Big data, and IoT) and those technologies affecting the production processes (innovative and collaborative robots and additive manufacturing), marginally the rest of other technologies. In addition, Fig. 3 shows also the intensity of investment in Industry 4.0 technologies. Results stress that Industry 4.0 is not a "single technology adoption" strategy but a technological "system" that needs more technologies as already found in the literature (Dalmarco, Ramalho, Barros, & Soares, 2019; Frank, Dalenogare, & Ayala, 2019). The most part of sample (70.5%) adopted at least three (median) technologies.

3.4 Industry 4.0 and KM in Top Performers: Survey Results

To evaluate the role and the value of Industry 4.0 for KM process of companies with the survey, we aimed to assess some strategic variables in order to define the relationship between Industry 4.0 and KM. In particular, Table 2 shows the motivation of adoptions and the impacts of new technologies on business listed in terms of importance.

In addition to the production efficiency (73.5%), which represents one of the first and most important antecedents of Industry 4.0 implementation (Kiel, Arnold, & Voigt, 2017), the other most relevant motivation of adoption refers to broader goals: creating new knowledge through market data and interactions with customers and other business partners (Smith, Collins, & Clark, 2005). In particular, *Champions* adopt the new technologies to improve customer service (74.4%), face the international competitiveness (72.3%), and try to exploit new marketing opportunities (50.0%), in terms of new market and new products development. Effectively, through Industry 4.0 technologies companies achieved improvements in the production (efficiency and productivity, respectively, 76.5% and 67.7%) and market (customer service and international competitiveness, respectively, 67.7% and 56.9%) sphere.

To explore the relationship between Industry 4.0 and KM, we also examined the impacts of new technologies on product and on working method/skills. Figure 4 shows how the new technologies affect the product offered by companies. Firstly, *Champions* use the new technologies to get higher control over product use (45.5%). In this way, they can get the data useful to improve production and marketing

Table 2 Motivations of adoption and impacts on business

Motivations of adoption	Frequency (%)	Impacts of I4.0 technologies	Frequency (%)
Improving customer service	74.4	Production costs efficiency	76.5
Production efficiency seeking	73.5	Higher productivity	66.7
Facing international competitiveness	72.3	Improved customer service	66.7
New marketing opportunities	50.0	Keeping international competitiveness	56.9
Improving environmental sustainability	39.0	Increased turnover	43.1
Enhancing product diversification	32.5	Higher product diversification	22.5
Requests from customers	27.5	New markets development	19.6
Maintaining production in Italy	25.6	Improved customized products share	19.6
Standard sector upgrading	20.0	Environmental sustainability	19.6
Imitating competitors	9.8	Relocalization of production activities	3.9
Reshoring-Backshoring	2.7		

$N = 61$

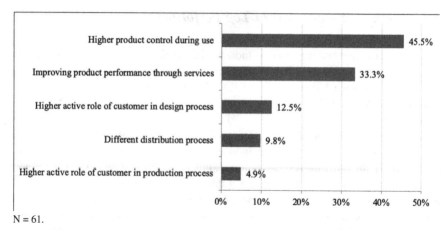

Fig. 4 Impacts of new technologies on product

performance. Secondly, they use the new technologies to improve product performance offering new services (33.3%). The collaboration with customers has a marginal value with respect to both the design (12.5%) and the production (4.9%) processes.

The new technologies have also a key role in internal working methods and relationships along value chain, as well as in job skills and competences (Arnold, Kiel, & Voigt, 2016; Nagy et al., 2018). As far as the impacts on production activities are concerned as well as relationships among company departments and with suppliers, Fig. 5 highlights interesting results for the research purposes.

Results reported in Fig. 6 stress the role of knowledge and its value in the Industry 4.0 paradigm. New knowledge creation for both product (41.9%) and production (40.9%) activities improvements is the main output the new technologies use and this depends on data that they are able to generate (Lu & Weng, 2018). The use of new technologies influences also the upgrading of skills and competences (26.7%) and the collaboration among the different business areas (25.0%). It is interesting to see that there is no reduction of human–machine interface (only 2.3%) so that technologies are not substituting completely workforce. Collaboration with suppliers has a marginal role in terms of impact of new technologies (11.4%).

Finally, we focused on skills and competences in terms of needs and changes (Fig. 6). The most important impact on employees' skills and competences refers to the technical area (62.0%), even if also administrative and managerial competences are interested from the Industry 4.0 revolution. These results confirm recent empirical research on the topic (Arnold et al., 2016; Whysall et al., 2019).

In this first exploratory part of the research, in addition to the analysis about Industry 4.0 adoption and its strategic impacts on business process in order to outline how the new technologies link to the KM process, we explored the key role of technologies for knowledge creation described in the theoretical section. Literature has shown that cloud, big data, IoT, and AI represent a group of Industry 4.0

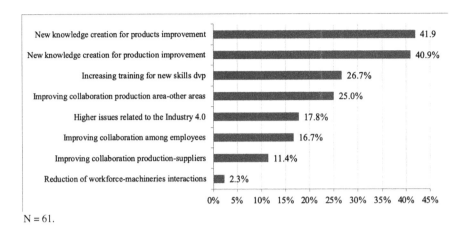

Fig. 5 Impacts of new technologies on internal and external working activities

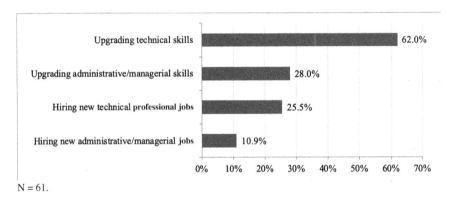

Fig. 6 Impacts of Industry 4.0 investments on skills and competences

technologies that more than others enable the gathering, storage, and management of data (Tao, Qi, Liu, & Kusiak, 2018; Xu, Frankwick, & Ramirez, 2016). Within our sample, Fig. 3 showed that cloud and big data are the most adopted technologies among those investigated. About 88% (54 on 61) of *Champions* adopted at least one of the four "data-driven technologies." This result stresses the importance of data management for the competitiveness of larger firms that compete at international level. Moreover, only 16.4% (10 on 61) of the sample adopted only such "data-driven" technologies, while most of the *Champions* invest also in other Industry 4.0 technologies. This result confirms the strong integration among the different technologies (Muscio & Ciffolilli, 2019).

As shown in Fig. 3, Cloud is the most adopted technology and big data is the second one. Big data became very important for larger firms because of the necessity to manage and analyze a remarkable amount of data gathered with the new technologies in production as well as in marketing and other business areas (Szalavetz,

Table 3 Correlations among data-driven technologies adopted

	Cloud	Big data	IoT	AI
Cloud	–			
Big data	−0.026	–		
IoT	0.138	0.095	–	
AI	0.224°	0.016	0.273*	–

$N = 1$; $* p < 0.05$; $° p < 0.10$

Table 4 Data-driven technologies integration

# Data-driven techs adopted	Cloud	Big data	IoT	AI
One	11 (28.2%)	9 (25.7%)	2 (9.1%)	0 (0.0%)
Two	15 (38.5)	15 (42.9%)	8 (36.4%)	0 (0.0%)
Three	11 (28.2%)	9 (25.7%)	10 (45.5%)	3 (60.0%)
Four	2 (5.1%)	2 (5.7%)	2 (9.1%)	2 (40.0%)
N	39 (63.9%[a])	35 (57.4%[a])	22 (36.1%[a])	5 (8.2%[a])

[a]% on the overall adopters (61)

2019). Moreover, *Champions* show a good IoT adoption rate; instead AI is the less adopted because of its early extensive use for business purposes (Haenlein & Kaplan, 2019). Performing a correlation analysis to explore the relationships among these four technologies adopted and how they correlate, it is interesting to observe (Table 3) that AI is the most correlated technology with the other ones and in particular with IoT ($0.273\ p < 0.05$) and Cloud ($0.224\ p < 0.10$).

Taking into consideration the adoption rate of these four technologies and the correlation values, it is interesting to see the growing integration of AI with other technologies as well as the role of big data as cross-sectional technology. Table 4 explores the *Champions'* strategies of investments in those four technologies, showing that AI investment is related to at least other two technologies, while IoT is the most integrated technology.

Data show that there is a sort of interdependency among the four data-driven technologies. Specifically, the integration of data-driven technologies, considering the sample adoption rate and the correlation among them, may be represented as shown in Fig. 7.

3.5 Data-Driven Technologies and Knowledge Management in Top Performers: Case Studies

Following the evidence emerged from quantitative analysis previously presented, we carried out a qualitative study aiming at understanding the relationship between Industry 4.0 and KM more deeply. Following recent research on Industry 4.0 (Müller 2019a, 2019b; Szalavetz, 2019; Vanchan, Mulhall, & Bryson, 2018), we focused the qualitative analysis, through a multiple case studies approach (Yin, 2009), on those companies that adopted mainly AI and other data-driven

Fig. 7 Integration among data-driven technologies

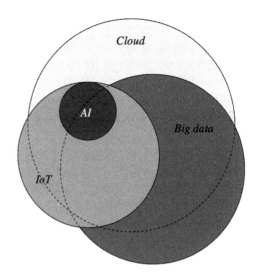

Table 5 *Champions* interviewed

Company	Activity	Industry 4.0 technologies adopted				
		Cloud	Big data	AI	IoT	Other technologies
1	Production of industrial lubricants	x	x	x		Manufacturing smart systems
2	Customized racing cars	x	x		x	• Manufacturing smart systems • Additive manufacturing • Laser cutting • 3D Scanner
3	Business systems solutions and packaging machinery	x	x	x	x	
4	Machinery for testing electronic products	x		x	x	
5	Professional smart kitchen ovens	x	x	x	x	• Manufacturing smart systems • Additive manufacturing • Laser cutting

technologies, in addition to the manufacturing-related Industry 4.0 technologies. We interviewed Chief Operations Officer or R&D managers of selected *Champions* companies, adopting AI solutions. Through the interviews with the COO or R&D managers of the five companies interviewed, we aimed at assessing the impact of new technologies on KM taking into account the use and the impacts on business processes and on workforce.

As Table 5 shows, Cloud is the common technology used by all firms interviewed. Moreover, most of them use data-driven technologies with other manufacturing-related technologies. In this way, we were able to explore the role of data and on knowledge produced by new technologies within both the business and manufacturing processes.

3.6 Results of Qualitative Analysis

Industry 4.0 investments and specifically data-driven technologies are related to KM strategies. We found that AI is mainly used to improve production processes as well as the product use by customers: data acquired—also in relation to IoT—allows the firm creating new knowledge.

> Artificial intelligence is used directly on production plants. It plays a key role on increasing the quality of the process, which transforms into better performance at the production level... (#1)

> Artificial Intelligence enables the centralization of operations management on the packaging lines and harmonizes the information coming from different sources in order to transform them from "Raw Data" to "Smart Data" (#3)

> The AI is used to acquire data, analyze them and exploit them in order to take decisions about some autonomous activities... For our business the data collected by AI have a very high value......the autonomous system must take the appropriate decisions on the basis of information of AI... (#4)

> Artificial Intelligence is mainly used in Business Intelligence area...the oven records and measures data that allow us to make diagnostics and maintenance in a predictive way and creating insights for customers to make the best use of the oven... (#5)

The analysis of interviews highlights the key role of data gathered through AI for the improvement of business processes through statistical machine learning and a consequent process of training and fitting models to data. The opportunity of gathering data through sensors and digitalization helps the firm in making a more in-depth analysis of the productions processes and use of the product. Those data are used by AI in order to find new association and a possible course of action that are validated by the operator or the manager.

In addition to the focus on specific technologies, companies need to reconsider their digital strategy (Davenport & Mahidhar, 2018). In particular, data become new knowledge that companies exploit to advance in operations management and deliver high-quality products on the market. Quality is one of the main competitive factors for companies as *Champions* that have a strong presence on the international markets. In addition to AI, *Champions* consider all the Industry 4.0 technologies very important for new knowledge creation.

> A company must continuously improve its knowledge and this drives us to invest constantly in new technologies (#2)

The oven has sensors that measure a whole series of things, the aggregate data arrive on Cloud where they are processed and getting insights that we provide to the customers (#5)

The first technology adopted was IoT. Now we are working with other technologies related to machine learning and deep learning (#5)

The interviews pointed out some interesting insights about how AI and also the other data-driven technologies enable companies to produce data and new knowledge that became very important for the sustaining of competitive advantage. Data and then knowledge allow firms improving the quality of production processes as well as of the product use, with direct consequences for the business growth. Data management, data analysis, and data mining enable knowledge-based decision-making processes (Brettel, Friederichsen, Keller, & Rosenberg, 2014). The new technologies allowed companies to achieve improvements in terms of both production quality and market and sales growth, especially by means of a faster new product development process, time to market response, and new services.

Artificial intelligence is used directly on production plants...increasing the stability, quality and the speed of the process...avoiding manual operations... (#1)

...the main result related to the use of new technologies is a stronger interest in our new products by our customers and potential customers... (#3)

Among the main results there are the increasing of prototyping and development processes and time to market response. Respect our business model...with the data we can offer additional services..." (#5)

Moreover, the new technologies allowed companies to improve relationships with suppliers and customers. In this way, companies were able to advance their supply chain activities and the customization process, with direct effects on customer satisfaction.

The introduction of Industry 4.0 technologies has increasingly connected the company with the suppliers, facilitating their quick feedback ... in the co-design process the technology is functional to improve collaboration processes, develop new technical ideas, simulate different scenarios and share technical experiences from different actors of the development process (#2)

We deliver on the market customized cognitive solutions and smart software characterized by self-learning systems, so customers can exploit the continuous learning process ... (#4)

The philosophy of the company is a very strong vertical integration ... when there is a new idea or a new technology we collaborate with the external environment ... integrating digital technologies into our processes (#5)

The research also aimed at understanding the relationship between the new technologies and the changes in terms of skills and competences, to consider how tacit and codified processes have to be integrated and the consequences in terms of human resources. The main goal was to verify the role of digital competences that are necessary for the success of Industry 4.0 implementation (Agostini & Filippini, 2019). The new knowledge (technical and managerial) that employees must have to manage data and processes (Butschan, Heidenreich, Weber, & Kraemer, 2019) becomes essential to compete effectively, being the lack of appropriate competences and skilled workforce one of the main barriers for the Industry 4.0 adoption (Horváth & Szabó, 2019). The interviews highlighted the importance of the technology and of the management (see Sect. 2.1)

> Artificial intelligence did not replace the operator but it is mainly a support for their work... Our skills were not sufficient...it was necessary to acquire new competences that we did not have...for the use of AI we were assisted by technology suppliers and external consultants (#1)

> The internal skills for the adoption of Industry 4.0 technologies are fundamental both in the identification phase and in the implementation phase.... The company is based fundamentally on research and development and innovation is its essence. To improve knowledge the culture of trial and error is promoted through the simulation of real phenomena. Moreover, we are a learning organization that facilitates the dissemination of knowledge and experiences. (#2)

> The main problem in the application of such an advanced technology is the lack of vertical skills to manage it fully ... so we needed of training ... The introduction of AI has certainly not had a negative impact on number of employees ... rather we had to hire and are still hiring new staff with advanced skills related to the AI. Prior investment in the Industry 4.0 allowed the development of a mindset that favours the introduction of more radical innovations as Artificial Intelligence" (#3)

> The new technologies shift the centre of gravity of the skills inside a company ... what the company had to do was create new know-how, guarantee training courses and create all the internal infrastructure to manage these types of projects. ... currently the company is looking for people who can use these innovative technologies ... the workforce, which in the past was 99% mechanical engineers, now consists of approximately the same number of mechanical engineers and other profiles who can exploit the technologies of the future. We talk in general about Data Scientist ... the main goal is to develop the digital competences and the know-how needed to manage the new technologies (#5)

The last verbatim of interviews highlights the key role of new competences for the successful implementation of AI and other Industry 4.0 technologies. In this case, the new knowledge is meant as new competences companies need to have inside if they aim to get the benefits form the implementation of AI and other Industry 4.0 technologies.

More generically, the Industry 4.0 paradigm bases its success on data and knowledge produced by the use of the new technologies. Companies should look for ways to incorporate that knowledge in their products and processes, as well as in a cognitive system able to integrate and share new knowledge created at the wider organizational level. The results of the analysis suggest the need to create a broader and better structured KM system related to this new industrial revolution. Then, this knowledge should be continuously improved and integrated with external partners and customers. The relationship between Industry 4.0 and KM seems, therefore, to be a strategic factor that might affect the competitive advantage of companies, more than what happened with the prior technological waves.

4   General Discussion and Conclusions

Industry 4.0 is a technological revolution that is shaping manufacturing and is changing how firms produce the product and how they interact in the value chain and with the consumer (Schwab, 2017). But as we discussed at the beginning of this chapter, it also a cognitive revolution. New technologies give the opportunity to create, store, and share new knowledge that plays an important role in reinforcing and/or developing firms' competitive advantage (Lu, 2017).

The results of the research show that firms' adoption process is more cautious, incremental, and longer than we could expect. Firms invested in ICT in order to define a sort of base layer of technological infrastructure on which to develop a more sophisticated application. From a KM perspective, firms adopted both an extractive and relational approach. As we saw from the results, firms adopted both ERP and software dedicated to the knowledge extraction as well as on the website, social media, and CRM that are dedicated to communication and interaction within and outside the firm.

The same seems to apply in the case of Industry 4.0. The firms are starting to invest in the more consolidated technologies available like cloud computing and are relatively less attracted by not well-established technology like AR or AI. This approach seems reasonable; firms are still learning how and when to adopt Industry 4.0. Nevertheless, if they have to start, they have to focus on the possibility to gather more data through cloud computing and big data. The main motivation that pushes firms to invest is the need for a better understanding of how the product is used by the consumer and on the production process. If they want to improve their product and the manufacturing process, firms need to have a clearer understanding of what is happening within and outside the factory.

In their process of adoption, firms seem to be driven by their business strategy than by a technological approach. Firms declare to have clearer strategic objectives that they want to reach like improving customer service, increasing production efficiency, and international competitiveness. In terms of KM, firms are interested in knowledge of the product and of the management and less keen on the technology side. As a matter of fact, they discover the need of knowledge of technology once they adopt and is remarkable that almost all the firms that invested in the technology declared the need of improving the technical skills within the company. This distance from the technological knowledge may explain the prudence with which they adopt Industry 4.0 solutions. Firms do not know how to use them properly; therefore, they opt for the ones that are more promising and in line with their strategic objectives. It is not surprising though that the firms that have already invested in several Industry 4.0 technologies are the ones which adopted more sophisticated and complex technologies like AI.

The qualitative analysis underlines the profound cognitive root of Industry 4.0. The firms that adopted AI are aware of both the new possibilities offered by this technology in order to analyze data and to propose a course of action and the importance of the judgment of a knowledgeable operator on the final decision.

This combination requires an increased amount of knowledge of the product and production processes (how could the product and process be improved?), of the management (how the data could be used for the firm?), and of the technology (how does AI work?). As the case studies pointed out, data are important but without the judgment of workers are not that useful. Tacit and explicit knowledge are strictly interconnected even in Industry 4.0 scenario.

References

Agostini, L., & Filippini, R. (2019). Organizational and managerial challenges in the path toward Industry 4.0. *European Journal of Innovation Management, 22*(3), 406–421.

Alavi, M., & Leidner, D. E. (2001). Review: Knowledge management and knowledge management systems: Conceptual foundations and research issues. *MIS Quarterly, 25*(1), 107–136.

Alcacer, V., & Cruz-Machado, V. (2019). Scanning the Industry 4.0: A literature review on technologies for manufacturing systems. *Engineering, Science and Technology, An International Journal, 22*, 899–919.

Ancarani, A., Di Mauro, C., & Mascali, F. (2019). Backshoring strategy and the adoption of Industry 4.0: Evidence from Europe. *Journal of World Business, 54*(4), 360–371.

Anderson, C. (2010). The next industrial revolution, atoms are the new bit, *Wired*, 01/15/2010.

Anderson, C. (2012). *Makers: The new industrial revolution.* New York: Crown Business.

Arnold, C., Kiel, D., & Voigt, K.-I. (2016). How the industrial internet of things changes business models in different manufacturing industries. *International Journal of Innovation Management, 20*(8). https://doi.org/10.1142/S1363919616400156.

Barney, J. B. (1991). Firm resources and sustained competitive advantage. *Journal of Management, 17*, 99.

Bharadwaj, A., El Sawy, O., Pavlou, P., & Venkatraman, N. (2013). Digital business strategy: Toward a next generation of insights. *MIS Quarterly, 37*(2), 471–482.

Bogers, M., Hadar, R., & Bilberg, A. (2016). Additive manufacturing for consumer-centric business models: Implications for supply chains in consumer goods manufacturing. *Technological Forecasting and Social Change, 102*, 225–239.

Bonfanti, A., Del Giudice, M., & Papa, A. (2018). Italian craft firms between digital manufacturing, open innovation, and servitization. *Journal of the Knowledge Economy, 9*(1), 136–149.

Brettel, M., Friederichsen, N., Keller, M., & Rosenberg, M. (2014). How virtualization, decentralization and network building change the manufacturing landscape: An Industry 4.0 perspective. *International Journal of Mechanical, Aerospace, Industrial and Mechatronics Engineering, 8*(1), 37–44.

Brown, J. S., & Duguid, P. (2000). *The social life of information.* Boston: Harvard Business Review Press. in 2017 the book was published with a new introduction.

Brown, J. S., & Duguid, P. (2001). Knowledge and organization: A social-practice perspective. *Organization Science, 12*(1), 198–213.

Butschan, J., Heidenreich, S., Weber, B., & Kraemer, T. (2019). Tackling hurdles to digital transformation – the role of competencies for successful industrial internet of things (IIoT) implementation. *International Journal of Innovation Management, 23*(4). https://doi.org/10.1142/S1363919619500361.

Chiarvesio, M., Di Maria, E., & Micelli, S. (2004). From local networks of SMEs to virtual districts: Evidence from recent trends in Italy. *Research Policy, 33*(10), 1509–1528.

Coreynen, W., Matthyssens, P., & Van Bockhaven, W. (2017). Boosting servitization through digitization: Pathways and dynamic resource configurations for manufacturers. *Industrial Marketing Management, 60*, 42–53.

Dalenogare, L. S., Benitez, G. B., Ayala, N. F., & Frank, A. G. (2018, July). The expected contribution of industry 4.0 technologies for industrial performance. *International Journal of Production Economics, 204*, 383–394. Elsevier B.V.

Dalmarco, G., Ramalho, F. R., Barros, A. C., & Soares, A. L. (2019). Providing industry 4.0 technologies: The case of a production technology cluster. *Journal of High Technology Management Research*. https://doi.org/10.1016/j.hitech.2019.100355.

Davenport, T. H. (2019). *The AI advantage. How to put the artificial intelligence revolution to work*. The MIT Press: Cambridge, MA.

Davenport, T. H., & Mahidhar, V. (2018). What's your cognitive strategy? *MIT Sloan Management Review, 59*(4), 18–23.

Davenport, T. H., & Prusak, L. (1998). *Working knowledge*. Boston: Harvard Business School Press.

Drucker, P. (1995). *"The post-capitalist executive," managing in a time of great change*. New York: Penguin.

Floridi, L. (2016). *The fourth revolution: How the Infosphere is reshaping human reality*. Oxford: Oxford University Press.

Frank, A. G., Dalenogare, L. S., & Ayala, N. F. (2019). Industry 4.0 technologies: Implementation patterns in manufacturing companies. *International Journal of Production Economics, 210*, 15–26.

Fry, H. (2018). *Hello world: Being human in the age of algorithms*. New York: W.W. Norton & Company.

Gershenfeld, N. (2012). How to make almost everything. *Foreign Affairs, 91*(6), 43–57.

Haenlein, M., & Kaplan, A. (2019). A brief history of artificial intelligence: On the past, present, and future of artificial intelligence. *California Management Review, 61*(4), 5–14.

Horváth, D., & Szabó, R. Z. (2019). Driving forces and barriers of industry 4.0: Do multinational and small and medium-sized companies have equal opportunities? *Technological Forecasting and Social Change, 146*, 119–132.

Kagermann, H., Helbig, J., Hellinger, A., & Wahlster, W. (2013). *Recommendations for Implementing the Strategic Initiative INDUSTRIE 4.0: Securing the Future of German Manufacturing Industry*. Final Report of the Industrie 4.0 Working Group. Forschungsunion.

Kamble, S., Gunasekaran, A., & Dhone, N. C. (2019). Industry 4.0 and lean manufacturing practices for sustainable organisational performance in Indian manufacturing companies. *International Journal of Production Research*. doi:https://doi.org/10.1080/00207543.2019.1630772.

Kane, G. C., Palmer, D., Phillips, A. N., Kiron, D., & Buckley, N. (2015). Strategy, not technology, drives digital transformation, becoming a digitally mature enterprise. *MIT Sloan Management Review*. http://sloanreview.mit.edu/projects/strategy-drives-digital-transformation.

Kiel, D., Arnold, C., & Voigt, K.-I. (2017). The influence of the industrial internet of things on business models of established manufacturing companies – A business level perspective. *Technovation, 68*, 4–19.

Kogut, B., & Zander, U. (1992). Knowledge of the firm, combinative capabilities, and the replication of technology. *Organization Science, 3*, 383–398.

Lasi, H., Fettke, P., Kemper, H.-G., Feld, T., & Hoffmann, M. (2014). Industry 4.0. *Business & Information Systems Engineering, 6*(4), 239–242.

Lee, J., Bagheri, B., & Kao, H. (2015). A cyber-physical systems architecture for industry 4.0-based manufacturing systems. *Manufacturing Letters, 3*, 18–23.

Leeflang, P. S., Verhoef, P. C., Dahlström, P., & Freundt, T. (2014). Challenges and solutions for marketing in a digital era. *European Management Journal., 32*(1), 1–12.

Lu, Y. (2017). Industry 4.0: A survey on technologies, applications and open research issues. *Journal of Industrial Information Integration, 6*, 1–10. Elsevier Inc.

Lu, H.-P., & Weng, C.-I. (2018). Smart manufacturing technology, market maturity analysis and technology roadmap in the computer and electronic product manufacturing industry. *Technological Forecasting and Social Change, 133*, 85–94.

McAadam, R., & Reird, R. (2001). SME and large organisation perception of knowledge management: Comparisons and contrasts. *Journal of Knowledge Management, 5*(3), 231–241.

Müller, J. M. (2019a). Antecedents to digital platform usage in industry 4.0 by established manufacturers. *Sustainability, 11*, 1121. https://doi.org/10.3390/su11041121.

Müller, J. M. (2019b). Business model innovation in small- and medium-sized enterprises: Strategies for industry 4.0 providers and users. *Journal of Manufacturing Technology Management.* https://doi.org/10.1108/JMTM-01-2018-0008.

Müller, J. M., Kiel, D., & Voigt, K.-I. (2018). What drives the implementation of Industry 4.0? The role of opportunities and challenges in the context of sustainability. *Sustainability, 10*, 247. https://doi.org/10.3390/su10010247.

Muscio, A., & Ciffolilli, A. (2019). What drives the capacity to integrate Industry 4.0 technologies? Evidence from European R&D projects. *Economics of Innovation and New Technology.* https://doi.org/10.1080/10438599.2019.1597413.

Nagy, J., Oláh, J., Erdei, E., Máté, D., & Popp, J. (2018). The role and impact of industry 4.0 and the internet of things on the business strategy of the value chain-the case of Hungary. *Sustainability (Switzerland), 10*(10). https://doi.org/10.3390/su10103491

Nonaka, I., & Takeuchi, H. (1995). *The knowledge-creating company: How Japanese companies create the dynamics of innovation.* Oxford: Oxford University Press.

Orr, J. (1996). *Talking about machines. An ethnography of a modern job.* ILR/Cornell University Press.

Pauleen, D. J. (2017). Davenport and Prusak on KM and big data/analytics: Interview with David J. Pauleen. *Journal of Knowledge Management, 21*(1), 7–11.

Pauleen, D. J., & Wang, W. Y. C. (2017). Does big data mean big knowledge? KM perspectives on big data and analytics. *Journal of Knowledge Management, 1*, 1–6.

Pine, B. J., II. (1993). *Mass customization. The new frontier in business competition.* Boston: Harvard Business Review Press.

Polanyi, M. (1966). *The tacit dimension.* Garden City, NY: Doubleday.

Porter, M. E., & Heppelmann, J. E. (2014). How smart, connected products are transforming competition. *Harvard Business Review, 92*(11), 64–88.

Schneider, P. (2018). Managerial challenges of industry 4.0: An empirically backed research agenda for a nascent field. *Review of Managerial Science, 12*(3), 803–848.

Schwab, K. (2017). *The fourth industrial revolution.* New York: Crown.

Smith, K. G., Collins, C. J., & Clark, K. D. (2005). Existing knowledge, knowledge creation capability, and the rate of new product introduction in high-technology firms. *Academy of Management Journal, 48*(2), 346–357.

Spender, J. C. (1996). making knowledge the basis of a dynamic theory of the firm. *Strategic Management Journal, 17*(special issues), 45–62.

Stentoft, J., & Rajkumar, C. (2019). The relevance of industry 4.0 and its relationship with moving manufacturing out, back and staying at home. *International Journal of Production Research.* doi. https://doi.org/10.1080/00207543.2019.1660823.

Sturgeon, J. S., Fredriksson, T., & Korka, D. (2017). *The 'new' digital economy and development.* UNCTD technical notes on ICT for development n.8.

Szalavetz, A. (2019). Industry 4.0 and capability development in manufacturing subsidiaries. *Technological Forecasting and Social Change, 145*, 384–395.

Tao, F., Qi, Q., Liu, A., & Kusiak, A. (2018). Data-driven smart manufacturing. *Journal of Manufacturing Systems, 48*(C), 157–169.

Tegmark, M. (2017). *Life 3.0: Being human in the age of artificial intelligence.* Knopf.

Tortorella, G. L., & Fettermann, D. (2018). Implementation of industry 4.0 and lean production in brazilian manufacturing companies. *International Journal of Production Research, 56*(8), 2975–2987. Taylor & Francis.

Vanchan, V., Mulhall, R., & Bryson, J. (2018). Repatriation or reshoring of manufacturing to the U.S. and UK: Dynamics and global production networks or from here to there and back again. *Growth and Change, 49*(1), 97–121.

Wang, L., Törngren, M., & Onori, M. (2015). Current status and advancement of cyber-physical systems in manufacturing. *Journal of Manufacturing Systems, 37*(2), 517–527.

Wang, S., Wan, J., Zhang, D., Li, D., & Zhang, C. (2016). Towards smart factory for industry 4.0: A self-organized multi-agent systems with big data based feedback and coordination. *Computer Networks, 101*, 158–168.

Whysall, Z., Owtram, M., & Brittain, S. (2019). The new talent management challenges of industry 4.0. *Journal of Management Development, 38*(2), 118–129.

Xu, Z., Frankwick, G. L., & Ramirez, E. (2016). Effects of big data analytics and traditional marketing analytics on new product success: A knowledge fusion perspective. *Journal of Business Research, 69*(5), 1562–1566.

Yao, X., & Lin, Y. (2016). Emerging manufacturing paradigm shifts for the incoming industrial revolution. *The International Journal of Advanced Manufacturing Technology, 85*(5–8), 1665–1676.

Yin, R. K. (2009). *Case study research: Design and methods*. Sage: Thousand Oaks, CA.

Zovico, F. (2018). *Nuove imprese. Chi sono i Champions che competono con le global companies*. Egea: Milan.

Zuboff, S. (2019). *The age of surveillance capitalism: The fight for a human future at the new frontier of power*. New York: PublicAffairs.

Industry 4.0 and Creative Industries: Exploring the Relationship Between Innovative Knowledge Management Practices and Performance of Innovative Startups in Italy

Silvia Blasi and Silvia Rita Sedita

Abstract The chapter explores how Industry 4.0 can affect knowledge management practices and innovation processes of companies operating in the creative industry. The analysis was carried out on a sample of 179 ICT (information and communications technologies) startups by administering a questionnaire. Through a cluster analysis we identify three types of creative, intensive, innovative startups: (1) laggards, (2) regular adopters, and (3) smart adopters. The laggards are characterized by a low adoption level of 4.0 technologies and a low turnover level; the regular adopters are the most technological; the smart adopters are the most economically performing cluster. Through further multivariate statistical elaborations, we characterize the three typologies on the basis of certain attributes, such as type, foundation year, and break-even point (BEP) reach. The results suggest that the strategic objective to get a round of financing can be considered the only differentiating factor in the market of ICT startups in Italy. These results underline the importance of funding for startups to serve their business objectives. Certainly, funding acts as the major constituents that support the growth of startups. The increasing level of competition led us to consider funding essential for matching the standards of the business world.

1 Introduction

The term *Industry 4.0* was introduced for the first time in Germany to define a policy framework aimed at fortifying the attractiveness of the German manufacture. This strategic initiative was adopted in November 2011 as part of High-Tech Strategy

---

S. Blasi (✉) · S. R. Sedita
Department of Economics and Management 'Marco Fanno', University of Padova, Padova, Italy
e-mail: silvia.blasi@unipd.it

2020[1] (Acatech, 2013). The goals of Industry 4.0 are to reach a higher working capacity level as well as a higher level of automatization (Thames & Schaefer, 2016). Industry 4.0 is characterized by a combination of several innovative technologies that is expected to significantly shift the landscape of the manufacturing industry. These technologies—advanced robotics, artificial intelligence, sophisticated sensors, cloud computing, and big data analytics—already exist in the manufacturing industry, but their integration facilitating interconnection and computerization into the traditional industry will transform this sector. The five major features of Industry 4.0 are (1) digitization, (2) optimization and the customization of production, (3) automation and adaptation, (4) human–machine interaction (HMI), and (5) value-added services and businesses and automatic data exchange and communication. These features are not only highly correlated with Internet technologies and advanced algorithms, but they also indicate that Industry 4.0 is an industrial process of value adding and knowledge management (Posada et al., 2015; Roblek, Meško, & Krapež, 2016).

In this chapter, we will explore how Industry 4.0 can affect knowledge management practices and innovation processes of companies operating in the creative industry. The increasingly sophisticated technological advancements can be used by creative industries to obtain up-to-date and relevant information based on a periodic and timely analysis process. Nevertheless, the academic literature does not reach a consensus about the effect of 4.0 technologies on creative industries. In this chapter, we focus on ICT (information and communications technologies) companies. The reason lies in the need to homogenize the sample composed of over 80% ICT startups. ICT industries are, in general, the suppliers of infrastructure, products, and services to the many ICT-using industries, which can be grouped into the telecommunications, ICT services, software, and hardware segments (Davis & Schaefer, 2003). Understanding their uses of 4.0 technologies and how they are related to their strategic orientation is crucial in order to start thinking about a technological ecosystem supportive of the 4.0 technologies adoption.

The analysis was carried out on a sample of 179 ICT startups by administering a questionnaire. A cluster analysis led to the identification of three types of creative, intensive, innovative startups. The first cluster is characterized by a low adoption level of 4.0 technologies and a low turnover level at the end of the third year. The second cluster is the most technical. The third is the most economically performing cluster. Through further multivariate statistical elaborations, we characterized the three typologies on the basis of certain attributes, such as type, foundation year, and break-even point (BEP).

---

[1]High-Tech Strategy was presented by the Federal Government in July 2010 and defines a generic process that needs to be effectively implemented for the targeted growth of the German research and innovation system.

2 Theoretical Framework

2.1 Creative Industries

In the recent literature, many studies explore the creative industries' contributions to the economy, particularly in terms of employment (Hearn, Bridgstock, Goldsmith, & Rodgers, 2014; Hesmondhalgh & Baker, 2013; Lazzeretti, Boix, & Capone, 2008), regional development (Andres & Chapain, 2013; Baum, O'Connor, & Yigitcanlar, 2008; Cooke & Lazzeretti, 2008; Oakley, 2004), and urban dynamics (Comunian, 2011; Landry, 2012; Levent, 2011).

Only recently has the academic literature investigated the role of innovation in creative industries in more detail. In particular, a group of studies focuses on innovation activities in creative industries (Galuk, Zen, Bittencourt, Mattos, & Menezes, 2016; Grimaldi & Grandi, 2005; Jones, Svejenova, Pedersen, & Townley, 2016; Lazzeretti, 2012; Müller, Rammer, & Trüby, 2009; Parkman, Holloway, & Sebastiao, 2012; Rice, 2002), while some other studies explore the role of creative industries as a driver of innovation, especially with regard to creative industries' inputs that may be used in other industries' innovation processes (Bakhshi & McVittie, 2009; Bakhshi, McVittie, & Simmie, 2008).

The creative industries are considered one of the most promising fields of economic activity in highly developed economies, having the potential to contribute to wealth and job creation. Their activities rest on individual creativity, skill, and talent. Nevertheless, their main output is intellectual property rather than material goods or immediately consumed services. Being a cross-sectional industry that operates with a large number of other sectors as well as public organizations and consumers, the creative industries profit from different customers and may encourage growth in a variety of other sectors by providing creative inputs. The creative industries may develop and foster innovations as part of their business activities, directly contributing an innovative output (products, services, technologies). Additionally, the creative industries support innovation in other industries through creative inputs (Lampel & Germain, 2016).

The creative industries' support for innovation in other sectors is strongly related to the concept of open innovation (Chesbrough, 2003; Enkel, Gassmann, & Chesbrough, 2009; Gassmann, Enkel, & Chesbrough, 2010; Laursen & Salter, 2006). Successful innovation often requires the combination of a firm's own innovative resources with external inputs, such as outside knowledge (new technology) or specialized research and development (R&D) services, to innovations generated by suppliers, competitors, or customers (Von Hippel, 2007, 2009). The creative industries, as producers of intellectual property, may be an attractive source of external knowledge for innovating firms. They offer a diverse variety of creative products and services, which can be integrated into other businesses' innovation processes, within an open knowledge management infrastructure. Furthermore, specific software can be developed to accomplish the needs of new products or processes, supporting the collection and diffusion of information useful to increase the production and marketing performance.

In the case of ICT companies, knowledge management practices are at the core of the most successful business models, where access to information from different sources is an important driver of competitive advantage.

2.2 Creative Industries and Industry 4.0

In this chapter, we explore creative, intensive, innovative startups, focusing on the digital ones. Specifically, we analyze these companies' strategic orientation toward Industry 4.0. Through Industry 4.0 or, in other terms, the "Fourth Industrial Revolution" (FIR), public and private institutions and literature (Edwards & Ramirez, 2016; Hirsch-Kreinsen, Weyer, & Wilkesmann, 2016; Lasi, Fettke, Kemper, Feld, & Hoffmann, 2014; Lu, 2017) refer to the evolution of the production of goods and services resulting from the application of new technologies. This transformation's key element is the intersection between production, processing processes and flows of information online (Internet of things, cloud, big data), and devices (sensors, chips) that communicate independently with one another along the entire value chain. To make strategically valuable use of 4.0 technologies, it is necessary to have an awareness of the strategic, organizational, and cultural levers they can enable. With such consciousness, suitable technologies can be chosen to create more business leverage and avoid low investment returns.

In recent decades, manufacturing and production systems have been gradually supplemented by information technology support instruments because controlling increasingly complex technologies, meeting the demands of multisite production, and supporting logistic processes have become even more complex tasks. In this context, ICT companies have come to represent images and expectations of the future, transforming both working conditions and efficiency (Harris, Wang, & Wang, 2015; Holtgrewe, 2014; Lyons, 2005). Industrial digitalization is trying to respond to rapidly changing customer needs: several new product variants expected by customers lead to a reduction of the product lifecycle. The technology related to the product's innovation should be based on the latest development. Nevertheless, a flexible production technology that can be transformed along with the ever-changing customer product specifications is required (Herrmann, Schmidt, Kurle, Blume, & Thiede, 2014). For manufacturing industries, industrial digitalization is leading to (1) a reduction in inventory, logistics, and material handling costs and (2) shorter lead times and fewer shortages during shipment (Biegler, Steinwender, Sala, Sihn, & Rocchi, 2018). Industry 4.0 is crucial for meeting these changing needs. For the implementation of Industry 4.0, tools that generate data and create big data are essential (i.e., 3D scanners, cameras, robots) (Zhong et al., 2015).

In Industry 4.0, startups will be able to make a difference. The process that will lead to automated and interconnected industrial production through the use of industrial analytics, the Internet of things, 3D printing, and cloud computing involves different players. An important role is certainly reserved for young innovative companies that can choose to run and offer their products or services independently or to become part of an open innovation path carried out by larger companies through partnerships or agreements. In Italy, the startups' centrality in the FIR is also reflected in the government's plan, which provides a series of measures in favor of new entrepreneurship: (1) tax deductions, (2) venture capital incentives, (3) the refinancing of the Central Guarantee Fund, and (4) the facilitation of the Ministry of Development Economic for small and medium-sized enterprises (SMEs) and startups. The plan could undergo further changes, but it is significant that it was designed with consideration of the startup ecosystem's development.

3 Methodology

A questionnaire was submitted between January 8, 2018, and March 14, 2018,[2] aimed at identifying the factors that lead to increasing the performance of innovative startups in creative industries. In particular, the survey aims to investigate (1) the role played by the company's relational network, (2) the company's main strategic objectives, and (3) its business model.

The questionnaire is composed of 24 items divided into three parts. The first section collects data about the company's principal characteristics (company name; activity; whether the startup is an independent company, a university spin-off, or a spin-off from an existing company; year of foundation; number of founders; startup phase (seed stage, startup stage, growth stage, later stage); number of employees; turnover; whether the company has reached the BEP and the main reference market). The second section refers to the principal collaborations, funding source, and 4.0 technologies adoption. Given the creative industries' growing economic importance, increased investment in innovation through digital content initiatives is the key to capturing future national benefits. Nevertheless, technologies have become commonplace and ubiquitous in the creative industries, often used as a means to directly enhance creativity, and in so doing, contribute to the life and culture of society as a whole as well as to identifying solutions to specific problems or ways to overcome barriers (Loveless, 2006). The third section explores the firm's strategic orientation, reflecting its operational, marketing, and entrepreneurial choice in an effort to enhance performance and gain a competitive advantage.

The questionnaires were collected through telephone interviews conducted by a Computer-Assisted Telephone Interview (CATI) service supply company. Starting with the universe composed of 2914 companies, the company contacted 1397 startups, realizing 219 interviews. The data processing took place through descriptive and multivariate analysis techniques using the statistical software STATA.

---

[2]This study is part of the Research University Project 2014 called "Moving knowledge into action: exploring the micro-foundation of an innovation ecosystem." It is an explorative analysis about the cultural and creative startups in Italy.

The startup types were identified through a hierarchical cluster analysis based on Ward distance (Everitt, 1979; Johnson, 1967). This made it possible to visualize the data structure through a dendrogram (or tree diagram), which facilitated the authors' choice regarding the number of groups to select (3). The cluster analysis is a set of statistical techniques designed to identify groups of units similar to one another in relation to a set of characteristics taken into consideration according to a specific condition. The objective is to combine heterogeneous units in several subsets that tend to be homogeneous and mutually exhaustive. The statistical units are, in other words, subdivided into a certain number of groups according to the level of "similarity" evaluated, starting from the values that a series of chosen variables assumes in each unit (Fabbris, 1990).

The classification variables used are the turnover at the end of the third year (euro) and the frequency of the use of 4.0 technologies, in particular of big data and the Internet of things.

We used the hierarchical grouping method that refers to the Ward algorithm, which suggests combining the two groups from whose fusion derives the minimum possible increase of the deviance "within" the groups at each stage of the aggregation process.

$$\text{DEV}_T = \sum_{s=1}^{p} \sum_{i=1}^{n} (x_{is} - \bar{x}_s)^2 = \sum_{i=1}^{n} \sum_{s=1}^{p} (x_{is} - \bar{x}_s)^2 \qquad (1)$$

where $\bar{x}_s$ is the mean of the variable s with reference to the whole collective. Given a partition in g groups, this deviation can be broken down into:

$$\text{DEV}_{\text{IN}} = \sum_{k=1}^{g} \sum_{s=1}^{p} \sum_{i=1}^{n_i} (x_{is} - \bar{x}_{s,k})^2 \qquad (2)$$

That is the deviation within the groups referred to the p variables with reference to group k, where $\bar{x}_{s,k}$ is the mean of the variable s with reference to group k.

$$\text{DEV}_{\text{OUT}} = \sum_{s=1}^{p} \sum_{k=1}^{g} (\bar{x}_{s,k} - \bar{x}_s)^2 n_k \qquad (3)$$

which is the deviance between the groups.

$$\text{DEV}_T = \text{DEV}_{\text{IN}} + \text{DEV}_{\text{OUT}} \qquad (4)$$

Going from groups $k+1$ to k (aggregation), DEV_{IN} increases, while DEV_{OUT} decreases. The Ward method joined those groups for which there is the least increase of deviance within the groups.

Table 1 Company distribution by ATECO macrosector

ATECO macrosector	Total	%
Other professional, scientific, and technical activities	11	5.02
Other manufacturing industries	1	0.46
Creative, artistic, and entertainment activities	3	1.37
Architectural and engineering activities; testing and technical analysis	1	0.46
Information service activities and other IT services	30	13.70
Film production, video and television programs, music and sound recordings	4	1.83
Publishing activities	16	7.31
Software production, IT consulting, and related activities	149	68.04
Advertising and market research	1	0.46
Printing and reproduction of recorded media	3	1.37
	219	

4 Sample

The universe is composed of 2914 startups operating in the cultural and creative industries included in the Italian register of the innovative companies born from 2012 to 2017. The sample includes 219 startups. Table 1 shows the ATECO distribution. Due to the large number of ICT companies, we decided to homogenize the sample, focusing on a subsample of 179 ICT companies.

The number of sampled startups per region is not equal; it changes according to the number of startup companies present in each region (proportional to size sampling). Table 2 shows the universe, sample, and subsample regional distribution.

Figure 1 illustrates each type of ICT startup for the foundation year. The subsample is predominantly composed of independent companies (85%), followed by academic spin-offs (8%) and corporate spin-offs (7%). The majority of the startups were established in the period between 2013 and 2016.

Figure 2 shows, through box plots, the turnover at the end of the first, second, and third year of activity. Looking at Fig. 2, the long upper whisker means that startup turnover is varied among the most positive quartile groups. The median thicknesses for the three groups seem to be different, in particular the median increases during these 3 years. Some outliers are present in each period. In general, we can see a positive trend; in fact, turnover grows at the end of each year. However, the main question remains is to realize when these startups become cash-flow positive? This question can be answered looking at the moment the startup reaches the BEP, where a business turns from taking a loss to making a profit. The term is usually used to describe a startup firm that seeks to reach a point of profitability after an initial period of losses supported by investors. The BEP is vital to companies because, before this stage, they cannot focus on expanding their business. Of the sample, 55% have not yet reached the BEP. At the end of the first year of activity, 17% have reached it, with another 17% at the end of the second year, 9% at the end of the third year, and 2% after the third year.

Table 2 Universe and sample regional distribution

Region	Universe Frequency	Percentage	Sample Frequency	Percentage	Subsample (ICT) Frequency	Percentage
Abruzzo	52	1.8	2	0.9	2	1.12
Basilicata	22	0.8	–	–		
Calabria	65	2.2	6	2.7	5	2.79
Campania	208	7.1	8	3.7	6	3.35
Emilia Romagna	283	9.7	26	11.9	21	11.73
Friuli Venezia Giulia	49	1.7	4	1.8	3	1.68
Lazio	344	11.8	31	14.2	26	14.53
Liguria	54	1.9	4	1.8	4	2.23
Lombardy	792	27.2	56	25.6	47	26.26
Marche	98	3.4	5	2.3	3	1.68
Molise	9	0.3	–	–		
Piedmont	155	5.3	12	5.5	10	5.59
Puglia	127	4.4	8	3.7	6	3.35
Sardinia	68	2.3	8	3.7	7	3.91
Sicily	121	4.2	6	2.7	5	2.79
Tuscany	129	4.4	8	3.7	6	3.35
Trentino Alto Adige	78	2.7	10	4.6	10	5.59
Umbria	37	1.3	7	3.2	6	3.35
Valle D'Aosta	7	0.2	1	0.5		
Veneto	216	7.4	17	7.8	12	6.70
Total	2914	100.0	219	100.0	179	100.00

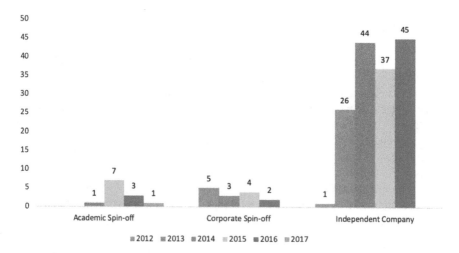

Fig. 1 Startup types and foundation years

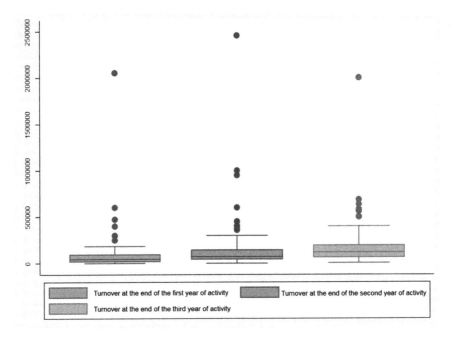

Fig. 2 Turnover at the end of the first, second, and third year of activity

Incubators assist emerging companies by providing support services and assistance in developing their businesses. They offer services structured with the final purpose of helping entrepreneurs overcome the difficulties they face when starting a new venture, with the hope of improving startup survival rates (Grimaldi & Grandi, 2005). Incubators differ from technological science parks, which can be defined as an area that allows an agglomeration of technological activities, leading to positive externality benefits for individual firms located within the park (Westhead, Batstone, & Martin, 2000). Almost 60% of the startups have decided to have no support in their seed stage. 38% passed through a private or public incubator and only the 3% through a technological science park.

The aim of public and private business incubators and technological parks is to remove much of the stress and strain from the startup, particularly with respect to creating the right support environment (Levitsky, 1991). Nevertheless, incubators and technological science parks (1) offer a fertile field for business (Martin, 1997), (2) provide and develop credibility for the new firm through admission into its special established environment and then from positive word of mouth through its own network and contacts (Rice, 2002), and (3) enhance client interactions that produce opportunities for business as firms use one another's services and pass on contacts (Hamdani, 2006). In Fig. 3, we can notice that 32 startups (18% of the sample), from those that were born or passed through an incubator or technological park, have reached the BEP. However, the majority has not yet reached it.

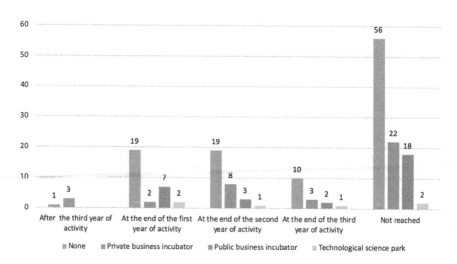

Fig. 3 Intersection between the break-even point and support in the startup seed stage

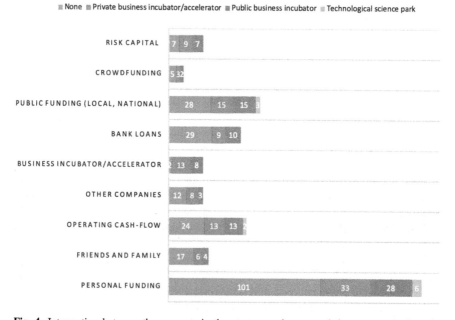

Fig. 4 Intersection between the supports in the startup seed stage and the company's financial sources

Figure 4 illustrates the intersection between the support received by startups in the seed stage and the company's financial sources. Starting and growing a new business require considerable financial resources. Besides the normal financial requirements that any new company faces, an ICT company needs additional money for its R&D

to develop new products and markets. Lack of financial resources is one of the major problems these companies face (Moore, 1993). From Fig. 4, it is possible to see that the entrepreneurs' personal funding is the most common financial source. Public funding, operating cash flow, and bank loans are used to raise funding.

The type of support the startup received in their seed stage did not influence its financial sources. Incubators and technological parks do not seem to influence this area. A better financial strategy is desirable.

5 Laggards, Regular Adopters, and Smart Adopters: Results from a Cluster Analysis

We performed a cluster analysis in order to classify firms by two relevant dimensions: turnover and 4.0 technologies. We isolated the two 4.0 technologies that can affect the business of ICT companies: big data and Internet of things. One of the most fundamental processes in a successful organization is the one involving transformation of information into knowledge resource. Big data and Internet of things offer best potential for supporting the aforementioned task.

We therefore consider two variables:

- turnover_3: Turnover at the end of the third year (euro)
- bigIoT_2 = bigdata + internet_things: frequency of the adoption of these two 4.0 technologies

As illustrated in Paragraph 2, Ward's hierarchical clustering method was applied; from the analysis of the dendrogram, we identified three groups of companies. Therefore, the sample of companies is divided into three clusters. Figure 5 represents

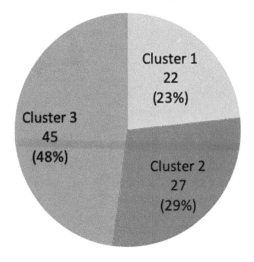

Fig. 5 Frequency distribution of the startups in three clusters

Table 3 Turnover at the end of the third year (thousand €) and 4.0 technologies adoption into the three clusters

		Turnover_3	bigIoT_2	
Cluster 1 Laggards	n.	22	22	
	Min	8.517393	0	
	Mean	10.04536	0.4545455	
	Max	11.28979	1	
Cluster 2 Regular Adopters	n.	27	27	
	Min	10.59666	1	
	Mean	11.37491	1	
	Max	11.73608	1	
Cluster 3 Smart Adopters	n.	45	45	
	Min	11.84941	0	
	Mean	12.48031	0.7777778	
	Max	14.50866	1	
Total	n.	94	179	
	Min	8.517393	0	
	Mean	11.59292	0.7486034	
	Max	14.50866	1	

the startups' frequency distribution in 3 clusters. We can see that the first cluster has 23% of the sample, the second has 29%, and the third has 48% (Table 3).

Cluster 1 (laggards) is the least numerous (22 startups) and is characterized by a turnover at the end of the third year and a frequency of the adoption of big data and the Internet of things below the total average. Cluster 1 is the least performing cluster; we can assume that it is composed of startups that have yet to establish themselves in the market and have a small role compared to their competitors. In addition, low turnover levels have also limited investments in 4.0 technologies.

Cluster 2 (regular adopters) is the most innovative. It is composed of 27 startups characterized by a high frequency of the adoption of 4.0 technologies: the minimum value is equal to the maximum (1). The average turnover at the end of the third year is just below the total average. The startups in this cluster, therefore, are in a growth stage, have invested in new technologies, and are gradually establishing themselves in the market. The extensive adoption of 4.0 technologies represents a cost that will impact on the economic performance in the long run.

Cluster 3 (smart adopters) is the most numerous (94 startups) and most performing in terms of average turnover, with an average turnover at the end of the third year over the total turnover. The same can also be said of the frequency of the 4.0 technologies adoption, where the cluster average is 0.77 with respect to the total average, which is equal to 0.74. Startups belonging to this cluster seem more selective in the choice of 4.0 technologies, adopting only the technologies that are

Industry 4.0 and Creative Industries: Exploring the Relationship Between... 125

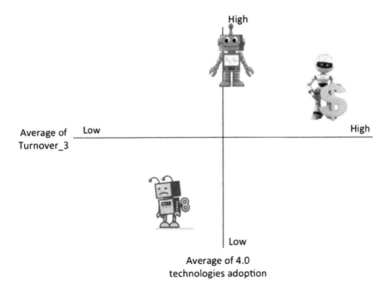

Fig. 6 Cluster matrix

Table 4 Type of the ICT startups for each cluster

Type	Cluster 1 *Laggards*	Cluster 2 *Regular Adopters*	Cluster 3 *Smart Adopters*
Corporate spin-off		2	2
Academic spin-off	2	2	3
Independent company	20	23	40
Total	22	27	45

strictly related to an increase of the economic performance. The startups in this cluster have established themselves in the market as compared to their competitors.

Figure 6 represents the cluster matrix, which shows the positioning of the three clusters of startups alongside the two dimensions of turnover and 4.0 technologies.

Having identified the three clusters of startups with reference to the segmentation variables, we will now define the competitive attributes that differentiate and characterize the three groups within the market.

Table 4 illustrates the types of ICT startups for each cluster. We can see that the three startup types of startups are distributed equally within the sample. Independent companies are the most numerous, while corporate spin-offs and academic spin-offs are residual. Looking at the foundation year of the ICT startups for each cluster, we observe that in Cluster 1 the majority of firms were born in 2014; in Cluster 2 between 2014 and 2015; finally, in Cluster 3 between 2013 and 2014. Therefore, Cluster 3 collects the oldest startups.

Fig. 7 BEP for each cluster

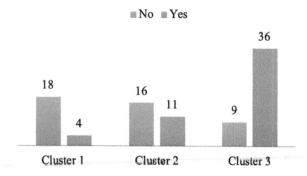

Figure 7 shows how many ICT startups for each cluster have reached the BEP. The majorities are in Cluster 3, where 36 ICT startups have reached the BEP, followed by Cluster 2 (11 startups) and Cluster 1 (4 startups).

6 Strategic Orientation and Financial Sources by Cluster

However, what is the strategic orientation toward Industry 4.0 of creative, intensive, innovative startups in Italy? To answer to this question, we did a factor analysis of the strategic orientation item, and then we used the analysis of variance (ANOVA) as a comparison method. Table 5 summarizes the variable through which we measured the strategic orientation. All are dummy variables (0,1).

Table 6 represents the factor analysis of the strategic orientation. As shown, there are 3 factors:

- Factor 1, which contains the variables "communication development" and "increase the brand awareness." We will rename this factor "Sales."
- Factor 2, which contains the variables "become a member of a cluster/business ecosystem," "develop partnerships with other startups in the same sector," and "develop partnerships with other startups in another sector." We will rename this factor "Collaboration with startups."
- Factor 3, which contains the variable "get a round of financing." We will rename this factor "Financing."

We then used the ANOVA[3] on the identified factors as a comparison method. The ANOVA method aims to verify the null hypothesis that there are no differences due to the analysis of treatments result (Testing H_0 against H_1).

---

[3]An ANOVA is a set of statistical techniques that are part of the inferential statistics that allow two or more groups of data to be compared by comparing the internal variability in these groups with the variability between the groups.

Table 5 Strategic orientation variables

Variable	Description
Stra_employees	Rapid growth in terms of number of employees
Stra_turnover	Rapid growth in terms of revenues
Stra_clients	Rapid growth in terms of number of clients
Stra_profit	Profitability growth
Stra_prodev	Product/service development
Stra_comm	Communication development
Stra_tech	Technology development
Stra_brand	Increase the brand awareness
Stra_inter	Internationalization
Stra_cluster	Become a member of a cluster/business ecosystem
Stra_samestartup	Develop partnerships with other startups in the same sector
Stra_diffstartup	Develop partnerships with other startups in another sector
Stra_round	Get a round of financing
Stra_othercoll	Other partnerships

Table 6 Factor analysis

Variable	Factor1 (Sales)	Factor2 (Collaboration with startups)	Factor3 (Financing)	Uniqueness
Stra_employees				0.8187
Stra_turnover				0.6056
Stra_clients				0.7089
Stra_profit				0.6483
Stra_prodev				0.6628
Stra_comm	0.5556			0.5411
Stra_tech				0.6968
Stra_brand	0.5761			0.4692
Stra_inter				0.7671
Stra_cluster		0.5371		0.6336
Stra_samestartup		0.6631		0.5261
Stra_diffstartup		0.6500		0.5155
Stra_round			0.5119	0.6710
Stra_othercoll				0.8057

$$H_0 : \mu_1 = \mu_2 = \ldots = \mu_p$$

H_1 : at least two means are different from each other

Through the ANOVA we worked on the composition of the variance. In particular, the variance can be distributed in two parts:

- The Variance Between (variance between groups): attributable to the treatment
- The Variance Within (variance within groups): residual within the groups

Table 7 Variance analysis output

Source of variation	df	Sum of square	Mean of square
Between treatment	$k-1$	S_A	$S_A/(k-1)$
Residual error	$N-k$	S_R	$S_R/(N-k)$
Total	$N-1$		

The total variance = the variance between groups + the residual variance

- The total variance = $\sum\sum (y_{ik}-y)^2/N-1$,
- The variance between groups = $n \sum (y_k-y)^2/K-1$,
- The residual variance = $\sum\sum (y_{ik}-y_k)^2/N-k$,

where $N-1$ is the total degrees of freedom, $K-1$ is the degrees of freedom "between treatment," and $N-k$ is the degrees of freedom of the residual error.

In conclusion, Table 7 results:

If the null hypothesis is true, the two estimates of variance will have the same expected value for the population variance, so the ratio between the variance between the groups and the variance within the groups has the expected value of 1 if the null hypothesis is true; otherwise, it has a value >1.

In general, an F-test is used to compare the variances, making a ratio between the variance between and the variance within.

$$F_e = \frac{S_A/(k-1)}{S_R/(N-k)} \tag{5}$$

The one-tailed distribution is used, referring to the degrees of freedom for both previously calculated factors.

If Fe > Ftab, it can be asserted that at least two means are statistically different.

If we want to know which means are different, we use t tests between the pairs of means. Corrections on the significance level are applied because more comparisons are made on the same data. In particular, we apply the Bartlett test, the most used test to verify the equality of the k samples' variances.

Table 8 suggests accepting the null hypothesis of the equality of means.

Also in this case, we accept the null hypothesis of the equality of means (Tables 9 and 10).

It is interesting to note that the strategic objective to get a round of financing appears to be a differentiating factor in the ICT startup market in Italy. The most digital cluster (Cluster 2) shows a value of financing highest with respect to the other clusters, on average, revealing the importance of this strategic goal for the smart adopters. The least performing cluster (Cluster 1) has a negative mean. In this case, the ANOVA test allows us to define that the means are statistically different ($F > 1$, with a significance level of 0.0023).

Table 8 ANOVA test on the "Sales" variable

Summary of "Sales"

	Obs.	Mean	Std. Dev.	Min.	Max.
Cluster 1 *Laggards*	22	0.0036342	0.7707505	−1.998741	0.7278792
Cluster 2 *Regular Adopters*	27	0.2103979	0.6944288	−1.711505	1.034971
Cluster 3 *Smart Adopters*	45	0.0402858	0.7063084	−1.872306	0.8277286
Total	179	−8.06e-10	0.7439661	−1.999962	1.034971

Analysis of variance

Source	SS	df	MS	F	Prob > F
Between groups	0.658340008	2	0.329170004	0.64	0.5308
Within groups	46.9635449	91	0.516082911		
Total	47.6218849	93	0.512063278		

Bartlett's test for equal variances: chi2(2) = 0.2967 Prob > chi2 = 0.862

Table 9 ANOVA test on the variable "Collaboration with startups"

Summary of "Collaboration with startups"

	Obs.	Mean	Std. Dev.	Min.	Max.
Cluster 1 *Laggards*	22	0.0018768	0.7715301	−0.9909779	1.110508
Cluster 2 *Regular Adopters*	27	0.1244853	0.8195651	−1.114224	1.219142
Cluster 3 *Smart Adopters*	45	−0.1122497	0.8325078	−0.9884286	1.414679
Total	179	−3.93e-09	0.7865326	−1.114224	1.414679

Analysis of variance

Source	SS	df	MS	F	Prob > F
Between groups	0.956563415	2	0.478281707	0.72	0.4896
Within groups	60.459341	91	0.664388363		
Total	61.4159044	93	0.660386069		

Bartlett's test for equal variances: chi2(2) = 0.1610 Prob > chi2 = 0.923

Table 10 ANOVA test on the variable "Financing"

Summary of "Financing"

	Obs.	Mean	Std. Dev.	Min.	Max.
Cluster 1 *Laggards*	22	−0.2816536	0.696455	−1.796085	0.5834501
Cluster 2 *Regular Adopters*	27	0.2048704	0.3107339	−0.4349588	0.7387623
Cluster 3 *Smart Adopters*	45	0.1842414	0.5611623	−1.708918	0.8843403
Total	179	9.78e-10	0.6285498	−1.905015	1.221557

Analysis of variance

Source	SS	df	MS	F	Prob > F
Between groups	3.78731639	2	1.8936582	6.49	0.0023**
Within groups	26.5522237	91	0.291782678		
Total	30.3395401	93	0.326231613		

Bartlett's test for equal variances: chi2(2) = 14.4886 Prob > chi2 = 0.001
**$p < 0.05$

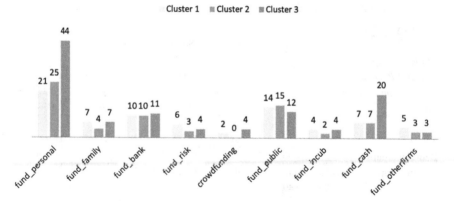

Fig. 8 Company's financial sources for each cluster

We decided to go more in depth with these results, analyzing which are the company's financial sources. The ICT companies could choose between:

- Personal funding
- Friends and family
- Bank loans
- Risk capital (venture/seed capital/business angel/private equity)
- Crowdfunding
- Public funding (local, national)
- Business incubator/accelerator
- Operating cash flow
- Other companies

Figure 8 illustrates the company's financial sources for each cluster. As shown, the most common financial source used is personal funding, followed by public funding and bank loans.

We create a new "funding" variable equal to the sum of the company's financial sources used and, as before, we applied the Bartlett's test on this new variable. Table 11 summarizes the results.

The ANOVA test allows us to refuse the null hypothesis of the equality of means according to the Bartlett's test which, in fact, reveals that the means are statistically different ($F > 1$, with a significance level of 0.0083).

We can notice that the laggards (Cluster 1) have the highest mean with respect to the other clusters. The regular adopters and smart adopters (Clusters 2 and 3) have similar means. This result can explain why, in Cluster 1, the strategic objective to get a round of financing has a negative value. As a matter of fact, startups that are willing to be at the frontier of the technological evolution are more sensitive to the source of financing. The propensity to invest on innovative and risky technologies is not a common feature of institutions like banks or public funding. Therefore, the regular and smart adopters can rely on a smaller portfolio of investors, thus urging to plan strategically the rounds of financing. The adoption of 4.0 technologies for creative

Table 11 ANOVA test on the variable "Funding"

Summary of "Funding"					
	Obs.	Mean	Std. Dev.	Min.	Max.
Cluster 1 *Laggards*	22	3.454545	1.534594	1	7
Cluster 2 *Regular Adopters*	27	2.555556	1.012739	1	4
Cluster 3 *Smart Adopters*	45	2.422222	1.287861	1	6
Total	179	2.430168	1.267301	1	7
	Analysis of variance				
Source	SS	df	MS	F	Prob > F
Between groups	16.5605846	2	8.28029228	5.05	0.0083**
Within groups	149.09899	91	1.63845044		
Total	165.659574	93	1.78128575		

Bartlett's test for equal variances: chi2(2) = 3.9589 Prob > chi2 = 0.138
**$p < 0.05$

industries is for sure an opportunity but at the same time represents a factor that increases the risk of failure. As well known from the literature and empirical evidence, innovators are often not supported until the market responds well to their offer. It is only afterward that investors recognize their value and are prone to sustain their activities. Consequently, the laggards enter the business created by the innovators, exploiting a successful path of development, which is created by regular and smart adopters.

7 Conclusion

This chapter investigates the strategic orientation toward Industry 4.0 of ICT startups in Italy. The cluster analysis is based on two relevant variables: the turnover at the end of the third year of activity and the adoption frequency of big data and the Internet of things. The results highlighted the existence of three clusters that we labeled laggards, regular adopters, and smart adopters.

The first cluster, the laggards, is characterized by a turnover at the end of the third year and an adoption frequency of big data and the Internet of things below the total average. The majority of these startups are born in 2014 and have not reached the BEP.

The second cluster, the regular adopters, is characterized by a high frequency of 4.0 technologies adoption and a turnover at the end of the third year just below the total average. The startups are born between 2014 and 2015, and almost half have reached the BEP.

The third cluster, the smart adopters, shows a turnover average at the end of the third year and an adoption frequency of 4.0 technologies over the total turnover. The startups are born between 2013 and 2015, and the majority has reached the BEP.

In search of a link between the heterogeneity of startups and their strategic goals, we discover that the strategic objective to get a round of financing can be considered the only differentiating factor in the market of ICT startups in Italy. The regular adopters have a larger capacity to attract financial resources, if compared to the others. The laggards are at disadvantage when trying to receive financial support, because they are not innovative and cannot guarantee a profitable investment. To further examine this aspect, we analyzed the financial sources of the startups. The laggards are the startups that use the largest portfolio of resources, while the regular and smart adopters have similar values. Investing in 4.0 technologies represents a great opportunity but also a great challenge for ICT startups, since the uncertainty linked to the potential benefits of this investments is a barrier for investors. A clearer view of the advantages of the adoption of 4.0 technologies should be provided, in order to solicit further uses of the technologies. Concerning the cross-fertilization between creative and manufacturing industries, we argue that traditional companies should be more prone to collaborate with ICT companies in order to create the fruitful environment needed to host and support the implementation of 4.0 technologies. A systemic view of the adoption of 4.0 technologies could encourage investors to sustain the development of businesses based on new tools of knowledge management, such as Internet of things and big data.

These results underline the importance of funding for startups to serve their business objectives. Certainly, funding acts as the major constituents that support the growth of startups. The increasing level of competition led us to consider funding essential for matching the standards of the business world. Therefore, funding activities improve the standards of business and contribute to business growth by enhancing the level of startups, according to the highest competition level in the corporate world, and should be utilized to stabilize startup businesses.

In fact, there are several financial tasks of an organization that need to be considered and managed. This task arrangement and management can be successfully attained when funds have been rightly allocated. This can be considered a crucial point, as the laggards, even if they received more funding than the regular and smart adopters, are the least performing and have an investment in 4.0 technologies below the total average.

Nevertheless, all startups need to ensure that their growth is free of hindrances by utilizing funding and fundraising programs and reducing personal funding. These programs, in fact, are aimed at raising and managing a startup's funds and, for this reason, should be monitored at regular intervals. In addition, fundraising programs are intended to remove all the financial blockages from the path of success for startups.

References

Acatech, F. (2013). Umsetzungsempfehlungen für das Zukunftsprojekt Industrie 4.0. *Abschlussbericht Des Arbeitskreises Industrie 4.0.*
Andres, L., & Chapain, C. (2013). The integration of cultural and creative industries into local and regional development strategies in Birmingham and Marseille: Towards an inclusive and collaborative governance? *Regional Studies, 47*(2), 161–182. https://doi.org/10.1080/00343404.2011.644531.

Bakhshi, H., & McVittie, E. (2009). Creative supply-chain linkages and innovation: Do the creative industries stimulate business innovation in the wider economy? *Innovations, 11*(2), 169–189. https://doi.org/10.5172/impp.11.2.169.

Bakhshi, H., McVittie, E., & Simmie, J. (2008). *Creating innovation. Do the creative industries support innovation in the wider economy?* NESTA Research Report March 2008, London.

Baum, S., O'Connor, K., & Yigitcanlar, T. (2008). The implications of creative industries for regional outcomes. *International Journal of Foresight and Innovation Policy, 5*(1–3), 44–64. https://doi.org/10.1504/IJFIP.2009.022098.

Biegler, C., Steinwender, A., Sala, A., Sihn, W., & Rocchi, V. (2018). Adoption of factory of the future technologies. In *2018 IEEE International Conference on Engineering, Technology and Innovation (ICE/ITMC)* (pp. 1–8). IEEE. doi:https://doi.org/10.1109/ICE.2018.8436310.

Chesbrough, H. W. (2003). *Open innovation: The new imperative for creating and profiting from technology*. Boston: Harvard Business School Press.

Comunian, R. (2011). Rethinking the creative city: The role of complexity, networks and interactions in the urban creative economy. *Urban Studies, 48*(6), 1157–1179. https://doi.org/10.1177/0042098010370626.

Cooke, P. N., & Lazzeretti, L. (2008). *Creative cities, cultural clusters and local economic development*. Cheltenham: Edward Elgar.

Davis, C. H., & Schaefer, N. V. (2003). Development dynamics of a startup innovation cluster: The ICT sector in New Brunswick. In D. Wolfe (Ed.), *Clusters old and new: The transition to a knowledge economy in Canada's regions* (pp. 121–160). Montreal: McGill-Queen's University Press.

Edwards, P., & Ramirez, P. (2016). When should workers embrace or resist new technology? *New Technology, Work and Employment, 31*(2), 99–113. https://doi.org/10.1111/ntwe.12067.

Enkel, E., Gassmann, O., & Chesbrough, H. (2009). Open R&D and open innovation: Exploring the phenomenon. *R&D Management, 39*(4), 311–316. https://doi.org/10.1111/j.1467-9310.2009.00570.x.

Everitt, B. S. (1979). Unresolved problems in cluster analysis. *Biometrics, 35*, 169–181. https://doi.org/10.2307/2529943.

Fabbris, L. (1990). *Analisi esplorativa di dati multidimensionali*. Padova: Cleup Editore.

Galuk, M. B., Zen, A. C., Bittencourt, B. A., Mattos, G., & Menezes, D. C. D. E. (2016). Innovation in creative economy micro-enterprises: A multiple case study. *Revista de Administração Mackenzie, 17*(5), 166–187. https://doi.org/10.1590/1678-69712016/administracao.v17n5p166-187.

Gassmann, O., Enkel, E., & Chesbrough, H. (2010). The future of open innovation. *R&D Management, 40*(3), 213–221. https://doi.org/10.1111/j.1467-9310.2010.00605.x.

Grimaldi, R., & Grandi, A. (2005). Business incubators and new venture creation: An assessment of incubating models. *Technovation, 25*(2), 111–121. https://doi.org/10.1016/S0166-4972(03)00076-2.

Hamdani, D. (2006). *Conceptualizing and measuring business incubation*. Ottawa: Statistics Canada, Science, Innovation and Electronic Information Division.

Harris, I., Wang, Y., & Wang, H. (2015). ICT in multimodal transport and technological trends: Unleashing potential for the future. *International Journal of Production Economics, 159*, 88–103. https://doi.org/10.1016/j.ijpe.2014.09.005.

Hearn, G., Bridgstock, R., Goldsmith, B., & Rodgers, J. (2014). *Creative work beyond the creative industries: Innovation, employment and education*. London: Edward Elgar.

Herrmann, C., Schmidt, C., Kurle, D., Blume, S., & Thiede, S. (2014). Sustainability in manufacturing and factories of the future. *International Journal of Precision Engineering and Manufacturing-Green Technology, 1*(4), 283–292. https://doi.org/10.1007/s40684-014-0034-z.

Hesmondhalgh, D., & Baker, S. (2013). *Creative labour: Media work in three cultural industries*. London: Routledge.

Hirsch-Kreinsen, H., Weyer, J., & Wilkesmann, J.P.D.M. (2016). *"Industry 4.0" as promising technology: Emergence, semantics and ambivalent character.* Sociological Working Papers, Technical University of Dortmund, Germany.

Holtgrewe, U. (2014). New technologies: The future and the present of work in information and communication technology. *New Technology, Work and Employment, 29*(1), 9–24. https://doi.org/10.1111/ntwe.12025.

Johnson, S. C. (1967). Hierarchical clustering schemes. *Psychometrika, 32*(3), 241–254. https://doi.org/10.1007/BF02289588.

Jones, C., Svejenova, S., Pedersen, J. S., & Townley, B. (2016). Misfits, mavericks and mainstreams: Drivers of innovation in the creative industries. *Organisation Science, 37*(6), 751–768. https://doi.org/10.1177/0170840616647671.

Lampel, J., & Germain, O. (2016). Creative industries as hubs of new organizational and business practices. *Journal of Business Research, 69*(7), 2327–2333. https://doi.org/10.1016/j.jbusres.2015.10.001.

Landry, C. (2012). *The creative city: A toolkit for urban innovators.* London: Earthscan.

Lasi, H., Fettke, P., Kemper, H.-G., Feld, T., & Hoffmann, M. (2014). Industry 4.0. *Business and Information Systems Engineering, 6*(4), 239–242. https://doi.org/10.1007/s12599-014-0334-4.

Laursen, K., & Salter, A. (2006). Open for innovation: The role of openness in explaining innovation performance among UK manufacturing firms. *Strategic Management Journal, 27*(2), 131–150. https://doi.org/10.1002/smj.507.

Lazzeretti, L. (2012). *Creative industries and innovation in Europe: Concepts, measures and comparative case studies.* Abingdon: Routledge.

Lazzeretti, L., Boix, R., & Capone, F. (2008). Do creative industries cluster? Mapping creative local production systems in Italy and Spain. *Industry and Innovation, 15*(5), 549–567. https://doi.org/10.1080/13662710802374161.

Levent, T. B. (2011). *Sustainable city and creativity: Promoting creative urban initiatives.* Farnham, Surrey: Ashgate Publishing.

Levitsky, J. (1991). Science parks and business incubators in the promotion of innovative SMI. *Small Enterprise Development, 2*(2), 47–51. https://doi.org/10.3362/0957-1329.1991.019.

Lu, Y. (2017). Industry 4.0: A survey on technologies, applications and open research issues. *Journal of Industrial Information Integration, 6*, 1–10. https://doi.org/10.1016/j.jii.2017.04.005.

Loveless, A. (2006). Professional development for technology and education: Barriers and enablers. *Technology, Pedagogy and Education, 15*(2), 139–141.

Lyons, M. H. (2005). Future ICT systems—understanding the business drivers. *BT Technology Journal, 23*(3), 11–23.

Martin, F. (1997). Business incubators and enterprise development: Neither tried or tested? *Journal of Small Business and Enterprise Development, 4*(1), 3–11. https://doi.org/10.1108/eb020975.

Moore, B. (1993). *Financial constraints to the growth and development of small high-technology firms.* University of Cambridge Small Business Research Centre Working Paper, 31 (July).

Müller, K., Rammer, C., & Trüby, J. (2009). The role of creative industries in industrial innovation. *Innovations, 11*(2), 148–168. https://doi.org/10.5172/impp.11.2.148.

Oakley, K. (2004). Not so cool Britannia: The role of the creative industries in economic development. *International Journal of Cultural Studies, 7*(1), 67–77. https://doi.org/10.1177/1367877904040606.

Parkman, I. D., Holloway, S. S., & Sebastiao, H. (2012). Creative industries: Aligning entrepreneurial orientation and innovation capacity. *Journal of Research in Marketing and Entrepreneurship, 14*(1), 95–114. https://doi.org/10.1108/14715201211246823.

Posada, J., Toro, C., Barandiaran, I., Oyarzun, D., Stricker, D., de Amicis, R., et al. (2015). Visual computing as a key enabling technology for Industrie 4.0 and industrial internet. *IEEE Computer Graphics and Applications, 35*(2), 26–40. https://doi.org/10.1109/MCG.2015.45.

Rice, M. P. (2002). Co-production of business assistance in business incubators: an exploratory study. *Journal of Business Venturing, 17*(2), 163–187. https://doi.org/10.1016/S0883-9026(00)00055-0.

Roblek, V., Meško, M., & Krapež, A. (2016). A complex view of industry 4.0. *Sage Open, 6*(2).

Thames, L., & Schaefer, D. (2016). Software-defined cloud manufacturing for industry 4.0. *Procedia CIRP, 52*, 12–17. https://doi.org/10.1016/j.procir.2016.07.041.

Von Hippel, E. (2007). Horizontal innovation networks—by and for users. *Industrial and Corporate Change, 16*(2), 293–315. https://doi.org/10.1093/icc/dtm005.

Von Hippel, E. (2009). Democratizing innovation: The evolving phenomenon of user innovation. *International Journal of Innovation Science, 1*(1), 29–40. https://doi.org/10.1260/175722209787951224.

Westhead, P., Batstone, S., & Martin, F. (2000). Technology-based firms located on science parks: The applicability of Bullock's 'soft-hard' model. *Enterprise and Innovation Management Studies, 1*(2), 107–139. https://doi.org/10.1080/14632440050119550.

Zhong, R. Y., Huang, G. Q., Lan, S., Dai, Q. Y., Chen, X., & Zhang, T. (2015). A big data approach for logistics trajectory discovery from RFID-enabled production data. *International Journal of Production Economics, 165*, 260–272. https://doi.org/10.1016/j.ijpe.2015.02.014.

Coordinating Knowledge Creation: A Systematic Literature Review on the Interplay Between Operational Excellence and Industry 4.0 Technologies

Toloue Miandar, Ambra Galeazzo, and Andrea Furlan

Abstract In the process of creating new knowledge, literature has scarcely studied how bodies of knowledge arising from different sources should be coordinated to enhance performance. In particular, the present research focuses on two sources of newly created knowledge, i.e., operational excellence and Industry 4.0, to understand whether they should be implemented sequentially or simultaneously. Operational excellence refers to the implementation of practices such as just in time, total quality management, and Six Sigma that help a firm to create knowledge that facilitates waste reduction and customer value improvement. Industry 4.0 refers to the implementation of new technologies such as artificial intelligence, big data, robotics, Internet of Things, and laser cutting that help a firm to create knowledge to improve overall business performance. We identified and analyzed 30 papers published in 13 peer-reviewed journals and conference proceedings in the field of operations management. Our findings based on the systematic literature review suggest that the interplay between operational excellence and Industry 4.0 can be categorized into four groups: (1) Industry 4.0 supports operational excellence; (2) operational excellence supports Industry 4.0; (3) complementary; and (4) no interdependence. Majority of the papers under study are in the first category, suggesting Industry 4.0 technologies as enabler of operational excellence.

1 Introduction

Organizational knowledge has increasingly been recognized as a central element of competitive advantage (Nonaka & Takeuchi, 1995). New knowledge is created as a result of a recursive interaction of tacit and explicit knowledge that continuously goes through four steps: socialization (from tacit to tacit knowledge), externalization

---

T. Miandar (✉) · A. Galeazzo · A. Furlan
Department of Economics and Management 'Marco Fanno', University of Padova, Padova, Italy
e-mail: toloue.miandar@unipd.it

(from tacit to explicit knowledge), combination (from explicit to explicit knowledge), and internalization (from explicit to tacit knowledge). Past literature has widely investigated the theory of knowledge creation across different disciplines, such as marketing, operations management, strategy, and innovation (Li, Huang, & Tsai, 2009; Linderman, Schroeder, Zaheer, Liedtke, Choo, 2004; Merx-Chermin & Nijhof, 2005; Moreno-Luzón & Begoña Lloria, 2008). Most of them, however, have focused on the effects that new knowledge has on a range of organizational processes leading to competitive advantage (Li et al., 2009; Tsai & Li, 2007). However, some scholars call for a better understanding of the process, not only the effects, of new knowledge creation (Gourlay, 2006).

Indeed, literature shows that the way activities are coordinated has profound implications on firm's performance (Barki & Pinsonneault, 2005; Galeazzo, Furlan, & Vinelli, 2014) and can greatly impact the way these activities are managed as well as their potential success. This chapter aims at contributing to the knowledge management literature by shedding some light on how it is possible to combine together different sources of knowledge. In particular, we focus on two different sources of knowledge creation: operational excellence and Industry 4.0.

On the one hand, operational excellence creates new knowledge by using a series of practices to eliminate each form of waste along the value chain. These practices, also known as lean practices or quality-related practices, increase the stability of processes by reducing machine setup times, guaranteeing overall equipment effectiveness, and introducing standard work. They also promote ways to create flow by replacing the push-oriented manufacturing planning and control systems with the adoption of a pull logic (Demeter & Matyusz, 2011) and to improve quality by eliminating scraps, defects, and reworks. Finally, these practices involve employees and increase their responsibilities and competences to sustain continuous improvement over time (Furlan & Vinelli, 2018; Galeazzo, Furlan, & Vinelli, 2017). Some of these practices are, for example, just in time (JIT), total quality management (TQM), total productive maintenance (TPM), human resource management (HRM), and Six Sigma (Galeazzo & Furlan, 2018; Furlan, Dal Pont, & Vinelli, 2011; Schroeder, Linderman, Liedtke, & Choo, 2008). Literature in operations management suggests that, as the implementation of these practices completely changes the way operators perform their jobs (e.g., they have a more in-depth understanding of the production processes, they are more involved in process improvements, and they collaborate more tightly with top management and colleagues), there is a strong relationship between operational excellence and the creation of new knowledge.

On the other hand, Industry 4.0 creates new knowledge because it represents a technological breakthrough for organizations and creates a paradigmatic change in the processes of value creation and competition rules. Industry 4.0 applied to manufacturing activities includes technologies such as additive manufacturing, advanced automation and advanced human–machine interface, Internet of Things—IoT, cloud manufacturing. These technologies have the potential to increase firms' efficiency and productivity, enabling them to strongly customize their products by flexibly adapting to the market demand (Holmström, Holweg, Khajavi, & Partanen, 2016; Roblek, Meško, & Krapež, 2016). Overall, literature on operations management agrees that technology, including Industry 4.0, allows operators to have access and incorporate explicit knowledge as well implicit knowledge as a result of the man–machine interaction. This implies that the interaction of tacit knowledge and explicit knowledge fosters knowledge creation.

Although literature is clear about the benefits of new technologies and operational excellence programs in creating new knowledge, the risk for firms is to approach Industry 4.0 and operational excellence as two separate cycles of knowledge creation. Firms implementing the two sources of knowledge independently risk reducing Industry 4.0 technologies to a mere technological investment, introducing new complexities, and digitalizing waste. Moreover, they risk operational excellence-related practices underperforming without an adequate technological support. It is important to combine the new knowledge created by Industry 4.0 and operational excellence in the most effective way. Thus, the idea that there are different ways to accrue the benefits of the combination between operational excellence and Industry 4.0 should be explored.

The purpose of this chapter is to provide a clear understanding of the combination of Industry 4.0 and operational excellence drawing on Thompson (1967)'s research on task coordination. In particular, there are three possible ways to coordinate operational excellence and Industry 4.0. First, they may be implemented separately having in mind that both of them contribute to the process improvement. Second, the joint implementation may occur sequentially and the main issue is to understand whether operational excellence should be implemented before or after the new technologies. Third, operational excellence and Industry 4.0 may be implemented together. Based on a systematic literature review, we found that only 30 papers deal with the purpose of this chapter, i.e., understanding the combination of Industry 4.0 and operational excellence. This is mainly due to the fact that research on Industry 4.0 is still at its infancy. Most of these studies argue that the introduction of Industry 4.0 technologies helps firms to exploit the potential of operational excellence (Roy, Mittag, & Baumeister, 2015; Rüttimann & Stöckli, 2016), thus implying that Industry 4.0 paves the way to the implementation of operational excellence. Some studies show that Industry 4.0 technologies need the support of operational excellence to maximize their potential in increasing performances (Khanchanapong et al., 2014; Rossini, Costa, Tortorella, & Portioli-Staudacher, 2019; Tortorella & Fettermann, 2018), thus implying that operational excellence paves the way to the implementation of Industry 4.0. Although contradictory, this evidence would suggest that the implementation of Industry 4.0 and operational excellence is sequential. Finally, few studies found that Industry 4.0 and operational excellence should be implemented simultaneously. Overall, these findings result in two important contributions. First, this chapter contributes to the literature on knowledge creation by providing practical examples of the way two sources of knowledge may be combined together. Second, it contributes to the literature on operations management by giving a state-of-the-art overview of the relationship between operational excellence and Industry 4.0.

2 The Process of Knowledge Creation

Though many researchers have been studying the process of creating knowledge, Dierkes, Antal, Child, and Nonaka (2003) identified the theory proposed by Nonaka (1994) as the stemming reference in knowledge creation literature. According to Google Scholar index, this paper has been cited more than 23,476 times whereas Scopus counted 320 citations, proving that Nonaka's theory has received an increasing attention since its publication. It has been described as a "highly respected" theory (Easterby-Smith & Lyles, 2003) and one of the most influential in knowledge management literature (Choo & Bontis, 2002). This theory has been applied to several areas of research as diverse as operations management (Galeazzo & Furlan, 2019; Linderman et al., 2004), innovation (Esterhuizen, Schutte, & Du Toit, 2012; Subramaniam & Youndt, 2005), human resource management (Droege & Hoobler, 2003), and internationalization strategies (Zahra, Ireland, & Hitt, 2000).

Nonaka (Nonaka, 1994; Nonaka & Takeuchi, 1995; Nonaka, Toyama, & Nagata, 2000) proposed that knowledge is created as a result of a continuous interaction of the epistemological and ontological dimensions of knowledge. The epistemological dimensions of knowledge comprise explicit and tacit knowledge. The former is easily accessible and codifiable because it refers to objective knowledge that is stored in such forms as documents, spreadsheets, standardized operating procedures, scientific formulas, and manuals. It is also easily shared among individuals within or outside the organization. The latter is difficult to classify, it resides in the know-hows of individuals, and it is linked to personal experience (Nonaka, 1994). The ontological dimensions of knowledge are classified as individual and social knowledge. Individual knowledge resides in individuals whereas social knowledge transcends individuals and it refers to knowledge that resides within groups, organizations, and even between organizations. The ontological dimensions represent the way knowledge can be disseminated throughout the different strata of an organization and transcend progressively beyond the boundaries of the organization.

Nonaka and Takeuchi (1995) depicted the process through which organizational knowledge is created by using a matrix, sometimes called the SECI model, which involves four sequential key activities of interaction between tacit and explicit knowledge: socialization (from tacit to tacit knowledge), externalization (from tacit to explicit knowledge), combination (from explicit to explicit knowledge), and internalization (from explicit to tacit knowledge). Through an iterative, spiral-like process, tacit knowledge is converted into explicit knowledge that, combined with new explicit knowledge, is finally internalized by the organization. This process does not stop once the activity of internalization has been performed, but continues by starting a new knowledge-creating spiral (see Fig. 1).

Socialization is the "process of sharing experiences and thereby creating tacit knowledge such as shared mental models and technical skills" (Nonaka & Takeuchi, 1995). This activity of knowledge interaction requires that individuals share their experiences and knowledge without the use of language through imitation, observation, and practice. Socialization is a time-consuming process because individuals are

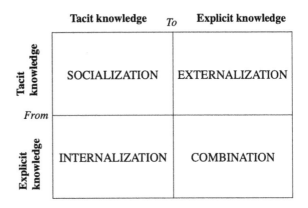

Fig. 1 The model of knowledge creation (adapted from Nonaka & Takeuchi, 1995, pp. 57, 62, 71)

supposed to spend time together, even through frequent physical proximity, and develop a relationship based on trust and empathy. Therefore, the core aspect of socialization is experience, as the mere transfer of information does not allow individuals to connect to each other to incorporate others' emotions and feelings and understand the specific context associated with the experience.

Externalization is the process of articulating tacit knowledge into explicit knowledge. "It is a quintessential knowledge creation process in that tacit knowledge becomes explicit, taking the shapes of metaphors, analogies, concepts, or models" (Nonaka & Takeuchi, 1995: 837). This activity of knowledge interaction requires that individuals communicate with one another through dialogue or collective reflection. In comparison to socialization, where knowledge is shared through unstructured and loosely defined interactions, externalization is often supported by structured and formal organizational mechanisms such as meeting and collaborative work assignments (Nonaka et al., 2000). The core aspects of externalization are language and symbols because they enable individuals to create mutually understandable knowledge. Therefore, externalization allows "the individually held tacit knowledge concepts to be crystallized and shared with other members, creating new knowledge" (Byosiere & Luethge, 2004: 246).

Combination is the process of combining different bulks of explicit knowledge. According to Nonaka and Takeuchi (1995), "reconfiguration of existing information through sorting, adding, combining, and categorizing of explicit knowledge can lead to new knowledge" (p. 67). Explicit knowledge is gathered from both inside and outside the organization and it is then disseminated among the employees of the organization. The use of technology can support the combination mode of knowledge creation as it facilitates the collection, synthesis, and dissemination of knowledge from different sources and its transformation into outputs such as reports, documents, and work rules that can be accessed from any part of the organization (Nonaka et al., 2000). Also the breakdown of knowledge can be considered a combination process. For example, breaking down corporate strategy into operational directions for the organization's functions is a way to create new explicit knowledge.

Internalization is the "process of embodying explicit knowledge into tacit knowledge" (Nonaka & Takeuchi, 1995). Through internalization, explicit knowledge created at the organizational level is internalized by employees, thus becoming tacit knowledge. Learning by doing, exercises, and training are different modes of knowledge internalization because they allow individuals to access newly created organizational knowledge and identify the knowledge important for themselves. "In practice, internalization relies on two dimensions. First, explicit knowledge has to be embodied in action and practice. [...] Second, there is a process of embodying the explicit knowledge by using simulations or experiments to trigger learning by doing processes" (Nonaka & Konno, 1998: 45).

Past research proves that the process of new knowledge creation has a prominent role in affecting performance and, thus, contributing to develop or sustain the firm's competitive advantage because knowledge is associated with innovative and difficult to imitate ways that enhance value creation for customers (Chang, Hung, & Lin, 2014; Jiang & Li, 2009; Tsai & Li, 2007). These studies mostly focused on assessing the effects of the combination of different sources of knowledge on performance. For example, Tsai and Li (2007) demonstrated that the implementation of new venture strategies triggers the dynamic spiral of knowledge creation that facilitates the successful execution of these strategies to improve performance. Similarly, Chang et al. (2014) provided empirical evidence on the positive relationship between knowledge creation, innovation, and creativity. Specifically, they found that knowledge creation enhances the ability of R&D personnel to develop products that include characteristics of novelty and that respond to customers' expectations, which in turn increases new product success. However, some scholars call for a better understanding of the process, not only the effects, of new knowledge creation (Gourlay, 2006). To fill this gap, this chapter investigates how knowledge is created by focusing on the context of operations management. In particular, the following sections will explore past literature to highlight the way two cycles of knowledge creation (new knowledge created by operational excellence and new knowledge created by Industry 4.0) are combined together.

3 Knowledge Creation in Operational Excellence

Firms are increasingly implementing operational excellence techniques like JIT, TQM, Six Sigma, and continuous improvement to reduce waste along the processes and enhance organizational performance. Literature highlights that knowledge creation and operational excellence are strongly connected. In fact, Deming (1994), one of the fathers of continuous improvement with his PDCA (Plan-Do-Check-Act) cycle, said that "best efforts and hard work, not guided by new knowledge, only dig deeper the pit we are already in" (p. 1). Moreover, Linderman et al. (2004) argued "organizations can create more knowledge by deploying quality management practices that support each of the knowledge creation processes (i.e., socialization, externalization, combination, internalization). Since knowledge creation often leads to improvement, effective deployment of quality management should result in a set of practices that support each of the knowledge creation processes" (pp. 601–602). Finally, Colurcio (2009) and Sin, Zailani, Iranmanesh, and Ramayah (2015) reported that there is an iterative interaction between operational excellence and knowledge creation because, on the one hand, operational excellence implements practices that facilitate the conversion of tacit knowledge into explicit knowledge to create new knowledge and, on the other hand, knowledge creation develops mechanisms that facilitate the adoption of operational excellence.

There are at least two reasons in support of the argument that the use of operational excellence techniques is strongly associated with knowledge creation. First, operational excellence aims at developing employees' systematic problem-solving behaviors to search for the root causes of problems and prevent errors to occur again. By adopting systematic problem-solving behaviors, employees contribute to the change of organizational routines. A routine "is a repetitive, recognizable pattern of interdependent actions, involving multiple actors" (Feldman & Pentland, 2003: 96) that consists of an ostensive (the schematic form of the routine) and performative (the actual way the routine is performed by individuals in a specific place and time) aspect. Furlan, Galeazzo, and Paggiaro (2019) argued that when employees analyze the causes of a problem, compare different alternatives to identify the most adequate solution, and, as a result, adopt actions, they modify the performative pattern of routines. These changes in organizational routines imply knowledge creation. Indeed, routines store knowledge that is embedded in organizational memory and, therefore, changes in routines modify, update, or revise existing knowledge. Likewise, Linderman, Schroeder, and Sanders (2010), drawing on the case study method, showed how Six Sigma enables knowledge creation. This study suggested that Six Sigma techniques enable employees to ask for the right questions and that, by getting the answers to the right questions, they created new knowledge. Moreover, newly created knowledge is shared among employees through organizational mechanisms such as meetings, teamwork, and standardized practices that, in turn, positively affect systematic problem-solving behaviors (Galeazzo & Furlan, 2019). It is therefore confirmed that systematic problem-solving behaviors, one of the most relevant micro-foundational elements of operational excellence, trigger activities of knowledge interaction and vice versa, thus reminding the knowledge creation process depicted by Nonaka and Takeuchi (1995).

Second, operational excellence fosters learning behaviors that, as an outcome, lead to the creation of new knowledge. Fine (1986) conducted one of the first researches on the relationship between quality improvement and learning. He found that the quality-based learning curve decreases manufacturing costs over time. This finding suggests that quality-based experience creates a better understanding of cost reduction. The author also found that cost reductions only depend on quality-based learning and not on other types of learning (i.e., autonomous learning and induced learning). The importance of learning behaviors that arise from the adoption of operational excellence is confirmed by Choo, Linderman, and Schroeder (2007). They showed that the adherence to structured methods linked to Six Sigma positively influences learning behaviors because structured methods define how to gather and process information in the most effective way. The development of learning behaviors, in turn, influences how information is interpreted and understood and it also shapes employees' thinking process, thus positively affecting knowledge creation. Moreover, Arumugam, Antony, and Kumar (2013) empirically demonstrated that operational excellence provides technical support to leaders for coordinating activities that foster learning in teams through the coordination of activities that transform individual knowledge into team-level knowledge that, as a result, enhances performance improvements. Therefore, literature shows evidence on the importance of operational excellence in generating new knowledge by promoting learning at individual, team, and organizational level that enables to reach the expected performance objectives, thus starting a new virtuous cycle of operational excellence, learning, and knowledge creation.

4 Knowledge Creation in Industry 4.0 Technologies

Technology is crucial to the success of any organization. Past literature on knowledge management has shown evidence of the relationship between technology and knowledge creation. In their paper focusing on the model of knowledge creation, Nonaka, Umemoto, and Senoo (1996) argued that "every business organization that wants to prosper in the knowledge society should fuse synergistically IT [i.e., technology] as knowledge-creation tools and human beings with collaborative knowledge creation capabilities to become a 'knowledge-creating company'" (p. 217). Technology improves the efficacy of knowledge-based processes as it gives access to knowledge, fosters knowledge sharing, facilitates collaboration among employees, enables the articulation and codification of knowledge, speeds up innovation processes, creates opportunities to combine different competencies and capabilities (e.g., using virtual environments), etc. (Arora & Gambardella, 1994; Santos, 2003; Vaccaro, Veloso, & Brusoni, 2009). Therefore, it is suggested that technology is a means through which knowledge creation flows.

The positive relationship between technology and knowledge creation is likely to be stronger in the era of Industry 4.0. The recent exponential development of Industry 4.0 technologies applied to manufacturing activities (i.e., additive manufacturing, advanced automation and advanced human–machine interface, Internet of Things—IoT, cloud manufacturing) leads technology to build a system of information and telecommunication technologies and industrial technologies that is more integrated and enables the operations function to become more information-led, digital, and responsive to customers compared to the past (Lee, Bagheri, & Kao, 2015). Industry 4.0 also changes the human–machine interaction because machines become increasingly autonomous and operators assume more responsibility, meaning their tasks are less related to mindless jobs. Instead, operators are asked to deal with a wide range of information that needs to be analyzed and take on the role of problem solvers to approach more complex problems (Gorecky, Schmitt, Loskyll, & Zühlke, 2014). Therefore, Industry 4.0 develops highly intelligent and interconnected factories in which highly skilled workers operate, suggesting there is an increasing opportunity for implementing cycles of new knowledge creation.

For a better understanding of the positive relationship between Industry 4.0 and knowledge creation, examples on the use of Internet of Things and artificial intelligence are provided. Related to IoT, this technology enables objects to upload data previously sensed into a central processing facility that, in turn, instructs objects to take actions, responding intelligently to changes in the environment. Hence, it provides information on productive assets and enables workers to quickly make adjustments in the most effective way in order to optimize production performance (Freedman, 2017). Compared to traditional factories in which sensors and devices have limited intelligence, the use of IoT ensures a more tight connection between physical and digital elements, a better communication and knowledge sharing, and decentralized decision-making processes. Thus, IoT enables two activities related to new knowledge creation: on the one hand, the transformation of tacit knowledge, embedded in objects, into explicit knowledge, embedded in information processing facilities, and, on the other hand, the internalization of explicit knowledge into tacit knowledge because workers have responsibility to make decisions based on the knowledge arisen from IoT technologies. Related to AR (augmented reality), this is a set of technologies that overlays digital data and images on the physical world, transforming volumes of data and analytics into images or animations. For example, wearable AR devices such as head-mounted displays or smart glasses allow workers to overlay digital information on real objects or environments (Porter & Heppelmann, 2017). Using wearable AR, they have easy access to instructions and detailed content about specific objects, materials, machines, or problems and, at the same time, they capture information and store them in the company's servers. This means that AR allows workers to process the physical and digital world simultaneously, rapidly and accurately absorbing information, making decisions, and executing required tasks quickly and efficiently. Thus, AR enables three activities of knowledge creation: first, it supports the externalization of tacit knowledge into explicit knowledge by allowing workers to integrate their personal knowledge with knowledge coming from digital data; second, it facilitates the combination of different chunks of explicit knowledge encompassed in the physical and digital world; and third, it promotes the internalization of explicit knowledge into tacit knowledge because workershave responsibility to make decisions.

5 The Coordination of Knowledge Creation

The implementation of operational excellence-related practices requires a coordinated set of work activities eventually performed by operators with the use of some tools and/or machineries. The implementation of digital technologies also requires that a coordinated set of work activities are performed. As seen in previous sections, both operational excellence and digital technologies use knowledge and generate additional knowledge. The main objective for the firm is to manage the work flows related to operational excellence and Industry 4.0 in order to assimilate and possibly combine the knowledge created by these two work flows in the most effective way to improve performance. In order to reach this objective, how to coordinate knowledge management emerges as a central topic that needs investigation. Knowledge coordination is defined as the management of interdependencies among work activities (Holsapple & Joshi, 2000). Therefore, it is important to understand the types of interdependencies between work activities.

Organizational literature and, in particular, theories on organization design have defined task interdependence in different ways (Galbraith, 1977; Hickson, Pugh, & Pheysey, 1969; Shea & Guzzo, 1987; Thompson, 1967; Van de Ven, Delbecq, & Koenig Jr, 1976). For example, Johnson and Johnson (1989) have distinguished task interdependence from resource interdependence, which is defined as the extent to which operators have to share the necessary resources. Shea and Guzzo (1987) argued that operators exercise discretion over task interdependencies that are viewed as an attribute of the operator. Other scholars (e.g., Van Der Vegt, Emans, & Van De Vliert, 2000, 2001) distinguished between goal interdependence and task interdependence. They stated that task interdependence depends on the way organizations design jobs and roles, thus affecting the division of information, materials, or expertise among operators. The degree of task interdependence typically increases as jobs become more complex and operators require others in order to reach the desired outcomes. For example, sales representatives operate almost independently from one another, whereas surgeons need great assistance from others to perform surgical operations. However, only Thompson (1967) explained how tasks can be designed to be executed at different levels of interdependence.

Thompson (1967) identified three patterns of task interdependence, each corresponding to a different degree of coordination between parts. Pooled interdependence is defined as a situation in which there is absence of workflow between parts. Each part acquires independent inputs and produces independent outputs that contribute to the whole organization. This implies that "each part renders a discrete contribution to the whole and each is supported by the whole" (p. 54). In this case, each part performs activities separately and in any order, without any exchange between parts. The second form of interdependence is defined as sequential interdependence, representing the situation in which each part's outputs are the inputs of another part, and similarly, the inputs that one part uses are the outputs from another part. This type of interdependence requires that parts perform activities in a specified sequence, implying there is an asymmetric exchange between parts. Finally, reciprocal interdependence represents a bidirectional exchange of inputs and outputs between parts, meaning that the activities performed by each part poses "contingency for the other" (p. 55).

The greater the interdependence, the more organizational decision-making is constrained through commitments, rules, and obligations, thus requiring higher coordination. With pooled interdependence, tasks are performed autonomously and coordination is achieved by standardization. This requires the implementation of routines and rules that define how and when each task should be performed and resources should be shared, thus minimizing the need for communication and decision-making among operators. With sequential interdependence, tasks are performed in sequence and coordination is achieved by plan. This implies the adoption of schedules for governing the workflow among operators. With reciprocal interdependence, multiple tasks are performed simultaneously and are strongly interconnected. In this case, coordination is achieved by mutual adjustment. Operators must continuously communicate with each other and give feedback in order to make adjustment whenever the expected objective becomes difficult to achieve.

Some scholars provided evidence of the existence of these types of task interdependencies in different areas of research. For example, Galeazzo et al. (2014) drew on case studies on pollution prevention to demonstrate that lean practices and environmental practices can be implemented either sequentially or reciprocally. Compared to a sequential interdependence, a reciprocal interdependence of lean and green practices leads to higher operational performance. Krishnan, Martin, and Noorderhaven (2006) used a survey on international strategic alliances operating in India in order to investigate the relationship between inter-organizational trust and performance when partners share tasks at different levels. They found that alliances benefit more from trust when partners show reciprocal interdependence rather than pooled or sequential interdependence. Gully, Incalcaterra, Joshi, and Beaubien (2002) studied the moderating effect of task interdependence on the relationship between team efficacy (i.e., a team's belief on its ability to successfully perform a specific task), potency (i.e., a team's belief on its ability to successfully perform any type of task), and performance. Their results indicated that team efficacy is more related to performance if task interdependence is high (reciprocal interdependence), whereas task interdependence did not moderate the relationship between potency and performance. Finally, Baumler (1971) provided evidence of the need to combine each type of interdependence with the respective coordination mode. For example, he showed that impersonal methods such as rules and routines were more frequently used with low task interdependence, and less frequently with high interdependence.

However, most areas of research have not highlighted the nature of task interdependencies between work flows related to different cycles of knowledge creation. To address this theoretical gap, this chapter focuses on the knowledge created by operational excellence and Industry 4.0 technologies. Understanding task interdependence in this context is particularly important because it is not clear whether firms implementing the two sources of knowledge independently may accrue higher or lower benefits than firms implementing operational excellence and Industry 4.0 simultaneously. Therefore, the purpose of this chapter is to review past studies related to both operational excellence and Industry 4.0 and examine them using the theoretical lens of Thompson (1967)'s research on task coordination.

6 Methodology

We have adopted a structured approach to the literature search and analysis in order to synthesize the results from previous research in the field. Given the large number of papers published about Industry 4.0 in the last few years, we deemed it necessary to adopt a systematic approach to identify and analyze the contributions that focused explicitly on operational excellence. Despite the large number of papers published on the emerging topic of Industry 4.0, there are few papers and specifically literature reviews addressing the interplay between operational excellence and Industry 4.0 technologies. Following Tranfield, Denyer, and Smart (2003), the key steps in a systematic review include the planning phase, the undertaking of the review, and reporting and dissemination. In order to address our research question, we conduct a systematic literature review identifying current state of academic research and contributions of the field (e.g., Schulze & Bals, 2018; Tranfield et al., 2003). This literature review systematically analyzes existing literature, examining publications on Industry 4.0 and operational excellence published in English, peer-reviewed journals and conference papers listed in the Scopus database. There is a lot of information in conferences especially on such emerging topics that are not published in journals yet; therefore, we included conference papers in our systematic literature review. The Scopus database was used because of its broader coverage. The literature review has been conducted in business, management, and accounting journals. The keywords that were used for searching in article title, abstract, and keywords fields were categorized into three groups:

- Industry 4.0/intelligent manufacturing/smart manufacturing
- Internet of Things/IoT/big data/artificial intelligence/AI/additive manufacturing/3D printing/cloud computing/collaborative robotics/augmented reality/virtual reality
- Lean management/lean manufacturing/lean thinking/operational excellence/six sigma/quality improvement/just in time/JIT/continuous improvement/total quality management/TQM/Kaizen.

Different combinations of these three groups were used to search for in the past literature in order to ensure that as many relevant articles as possible would be included. We tracked the papers until June 2019 when we conducted the search process.

The first stage of the search process generated 374 papers. The titles and abstracts of the papers within this initial sample were then checked manually for overall relevance. We removed duplicates and papers that were purely technical (e.g., about operational techniques in manufacturing) resulting in 60 potentially relevant papers. The relevant screening process of content of the papers, within the scope of relationship between operational excellence and Industry 4.0 technologies, further reduced the list to 30 relevant papers. We ended up (Fig. 2) with the 30 relevant papers from journals and conferences (Table 1) in the field that form the basis of our systematic literature review.

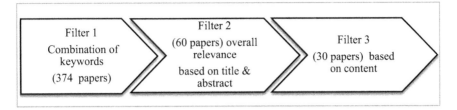

Fig. 2 Systematic literature review process

Table 1 Journals and number of papers identified for the final inclusion stage	International Journal of Production Research	10
	Conference paper	8
	Business Process Management Journal	1
	Central European Business Review	1
	IEEE Engineering Management Review	1
	International Journal of Product Development	1
	International Journal of Quality and Service Sciences	1
	Journal of Cleaner Production	1
	Journal of Industrial Engineering and Management	1
	Journal of Manufacturing Technology Management	1
	Journal of the Operational Research Society	1
	Systems Research and Behavioral Science	1
	Technovation	1
	Total Quality Management and Business Excellence	1

Table 1 shows the journals in which we identified relevant research papers and number of relevant papers in every journal. Highest number of papers is published in *International Journal of Production Research*, which is mainly reporting production and manufacturing research. Papers from conference proceedings are from International Conference of Industrial Engineering and Engineering Management, International Conference on Industrial Technology and Management, International Conference of Business Informatics Research, etc.

6.1 Data Analysis and Coding

We built an Excel database that contains data for all 30 papers. This step was the starting point in conducting the analysis presented in the next section. Regular meetings of three researchers to evaluate and finalize the analysis followed the coding. We commenced the analysis of the papers by examining methodology and year of publication.

We categorized the papers based on their focus into 4 groups: (1) Industry 4.0 supports lean manufacturing; (2) lean manufacturing supports Industry 4.0;

(3) complementary; and (4) no interdependence. Overall, our aim was to explore the relationship between Industry 4.0 and operational excellence.

7 Findings

In the following section, we provide a general overview of the results of the analysis as the basis for understanding the research approaches that have been applied in the field.

7.1 Number of Papers by Publication Year

Figure 3 shows that there is an increase in the number of papers regarding this topic, especially in the last few years. It is an emerging topic and together with published papers there are 8 relevant conference papers in the last few years. In our list there are a couple of papers published before the emergence of Industry 4.0 in literature, which were focused on specific technologies such as artificial intelligence (AI) and RFID (Brintrup, Ranasinghe, & McFarlane, 2010; Proudlove, Vadera, & Kobbacy, 1998).

7.2 Relationship Between Lean Manufacturing and Industry 4.0

Based on the focus of every paper, we tried to extract any kind of relationship they have addressed between Industry 4.0 and lean manufacturing. Accordingly we

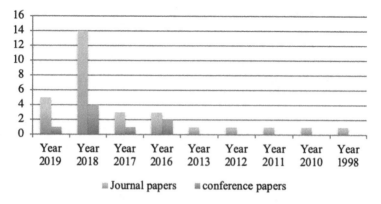

Fig. 3 Number of papers by publication year

Table 2 Number of papers addressing each category of relationship between Industry 4.0 and lean manufacturing

Category	Papers
Industry 4.0 supports lean manufacturing	14
Complementary	8
No interdependence	5
Lean manufacturing supports industry 4.0	3

allocated the papers under study into 4 categories; the first category refers to Industry 4.0 technologies as enabler of lean manufacturing, the second category refers to lean manufacturing lessons as enabler of Industry 4.0 improvement, the third category refers to simultaneous implementation of Industry 4.0 and lean manufacturing, and the fourth category is where importance of both Industry 4.0 technologies and lean manufacturing in order to achieve operational excellence has been acknowledged; however, there is no interdependence between the two. Table 2 shows the number of papers addressing each category.

We find the following for each of the four categories of papers; all papers under study in every category have been listed in Table 3.

1. *Industry 4.0 supports lean manufacturing* category is mainly regarding Industry 4.0 technologies that are supporting lean manufacturing. For example, RFID-enabled real-time traceability enhances the implementation of advanced strategies such as just-in-time (JIT) lean (Huang, Qu, Zhang, & Yang, 2012). RFID technology is viewed as a vehicle to achieve leaner manufacturing through automated data collection, assurance of data dependencies, and improvements in production and inventory visibility (Brintrup et al., 2010). AI technology based on big data is expected to promote the widespread use of quality management (Hyun Park, Seon Shin, Hyun Park, & Lee, 2017). Table 3 shows the list of journal and conference papers included in this category.

2. *Lean manufacturing supports Industry 4.0* category refers to papers such as Martinez (2019) stating Industry 4.0 is the next step after lean or other process improvement approaches. According to Beard-Gunter, Ellis, and Found (2019), there are positive implications merging good games design and TQM in socio-technic systems, which could improve engagement and quality in companies implementing in Industry 4.0.

3. *Complementary* meaning both lean management and Industry 4.0 support the objectives of operational excellence and they are stronger when being implemented together. Lean production practices are positively associated with Industry 4.0 technologies and their concurrent implementation leads to larger performance improvements (Tortorella, de Castro Fettermann, Frank, & Marodin, 2018). Not only can Industry 4.0 and lean thinking coexist but their integration can also provide benefits and opportunities (Demartini & Tonelli, 2018). Industry 4.0 needs to be understood as digitally enabled lean. Industry 4.0 solutions can enhance operational excellence, but they can also improve eco-efficiency (Szalavetz, 2017). The interaction between Industry 4.0 and lean manufacturing needs to be considered as two sides of operational excellence because the former's purpose is speeding up flows of information and the latter's

Table 3 List of papers included in each of the four categories of relationship between lean manufacturing and Industry 4.0

Category	Author(s)	Focus
Industry 4.0 supports lean manufacturing	Urbinati, Bogers, Chiesa, and Frattini (2019)	Quality management can be facilitated thanks to the quality records collected from the manufacturing processes
	Makhanya, Nel, and Pretorius (2018)—Conference paper	In benchmarking QM maturity traditional approach has reached the end of its lifespan, big data and impact of Industry 4.0 to be further investigated
	Bertoncel, Erenda, Bach, Roblek, Meško (2018)	Digitalization helps to detect early warning systems at smart factory, which is relevant to lean manufacturing
	Trotta and Garengo (2018)—Conference paper	"Committing into Industry 4.0 makes a factory lean besides being smart," while Industry 4.0 is being seen as the possibility to implement the lean automation in the factories
	Vogelsang, Packmohr, Liere-Netheler, and Hoppe (2018)—Conference paper	Industry 4.0 became part of the strategic orientation. Digital integration eases quality management and leads to higher demands
	Lugert, Batz, and Winkler (2018)	In general users appreciate a combination of lean methods and solutions of Industry 4.0. This paper provides a current evaluation of the VSM from an exploratory perspective
	Hyun Park et al. (2017)	AI supports the development and production of a high-quality product. AI technology based on big data is expected to promote the widespread use of QM
	Foidl and Felderer (2015)—Conference paper	Industry 4.0 provides promising opportunities for quality management Explore research challenges of Industry 4.0 for providing promising opportunities for QM
	Sanders, Elangeswaran, and Wulfsberg (2016)	Industry 4.0 is indeed capable of implementing lean; committing into Industry 4.0 makes a factory lean besides being smart
	Liu, Leat, Moizer, Megicks, and Kasturiratne (2013)	A knowledge system for lean supply chain management (KSLSCM) has been developed using artificial intelligence system shells VisiRule and Flex
	Huang et al. (2012)	RFID-enabled real-time traceability enhances the implementation of advanced strategies such as just-in-time (JIT) lean
	Zhang, Ong, and Nee (2011)	RFID in augmented reality environment—Aiming at providing just-in-time information rendering
	Brintrup et al. (2010)	RFID technology is viewed as a vehicle to achieve leaner manufacturing through automated data collection, assurance of data

(continued)

Table 3 (continued)

Category	Author(s)	Focus
		dependencies, and improvements in production and inventory visibility.
	Proudlove et al. (1998)	As a part of the LR: Contribution of AI techniques to both product or service quality management issues, and maintenance management
Lean manufacturing supports Industry 4.0	Beard-Gunter et al. (2019)	There are positive implications merging good games design and TQM in socio-technic systems which could improve engagement and quality in companies implementing in Industry 4.0
	Martinez (2019)	Industry 4.0 is the next step after lean or other process improvement approaches. It is about having the system coordinated
	Basios and Loucopoulos (2017)—Conference paper	The large amounts of data, especially in light of Industry 4.0, need to be organized so that organization can understand what insights they need in order to take strategic and operational decisions. In order to achieve that Six Sigma DMAIC Enhanced with Capability Modelling approach has been introduced
Complementary	Ren et al. (2019)	It has been estimated that the combination of big data analytics and lean management could be worth tens of billions of dollars, in improved profits for large manufacturers
	Tortorella and Fettermann (2018)	Lean production practices are positively associated with Industry 4.0 technologies and their concurrent implementation leads to larger performance improvements
	Buer, Strandhagen, and Chan (2018)	Industry 4.0 supports lean manufacturing through "Hard" practices, which refers to the technical and analytical practices used in lean like value stream mapping (VSM) and 3D printing
	Demartini and Tonelli (2018)—Conference paper	Not only can Industry 4.0 and lean thinking coexist but their integration can also provide benefits and opportunities
	Dallasega (2018)	Reorganization is required, preferably through lean management principles. Industry 4.0 concepts can support and foster the reorganization of processes. Variety of Industry 4.0 concepts to address these problems. For example, RFID technology allows gathering information about the construction supply chain (CSC) process in real time allowing for rapid response to unpredictable events

(continued)

Table 3 (continued)

Category	Author(s)	Focus
	Szalavetz (2017)	Industry 4.0 needs to be understood as digitally enabled lean. Industry 4.0 solutions can enhance operational excellence but they can also improve eco-efficiency, namely in the field of quality management (through smart production control, data analytics, and predictive modelling solutions); process optimization (through capacity planning and production scheduling solutions); and product and process engineering (through advanced virtual technologies).
	Eleftheriadis and Myklebust (2016)—Conference paper	Understanding best practices through TQM methods toward zero defect manufacturing. Then, thanks to the implemented sensors and monitoring systems, it is possible to provide a detailed documentation of any event occurred during the process named as Industry 4.0 or CPS
No interdependence	Hannola, Richter, Richter, and Stocker (2018)	This paper proposes a conceptual framework for empowering workers in industrial production environments with digitally facilitated knowledge management processes
	Yin et al. (2018)	The evolution of production systems from Industry 2.0 through Industry 4.0. Potential applications of lean principles for Industry 4.0 are presented
	Melnyk, Flynn, and Awaysheh (2018)	The best of times and the worst of times in operations and supply chain management (OSM)
	Chang and Yeh (2018)	About Industry 4.0 and the need for talent companies need talent in the areas of lean management, the Internet of Things (IoT), cloud computing, and big data
	Leyh, Martin, and Schäffer (2017)—Conference paper	Lean management/lean production principles are not often addressed in Industry 4.0 models. Despite the fact that those aspects are often seen as the basis for Industry 4.0 implementation this is not integrated in the respective models nor is it discussed in connection with these models

goal is eliminating waste to accelerate physical flows (Moeuf, Pellerin, Lamouri, Tamayo-Giraldo, Barbaray, 2018).

4. *No interdependence* category of papers is mainly literature reviews of the field, which do not necessarily focus on the relationship between Industry 4.0 and operational excellence; however, they touch upon both topics. For example, Yin, Stecke, and Li (2018) presented potential applications of lean principles for

Industry 4.0 in the evolution of production systems from Industry 2.0 through Industry 4.0. Leyh et al. (2017) analyze Industry 4.0 models with focus on lean production aspects.

7.3 Research Method

Figure 4 shows that the most frequent methods are literature review (LR), systematic literature review (SLR), and conceptual studies (with 16 papers). For example, the study by Hannola et al. (2018) identifies logical conclusions on empowering production workers with digitally facilitated knowledge processes in the form of a conceptual framework. More specifically, the study links current concepts, presenting new perceptions and expansion of the existing view (Gilson & Goldberg, 2015; Hannola et al., 2018), followed by case studies (with 8 papers). For example, Dallasega (2018) has collaborated with different engineers in order for supplier companies to optimize their processes using Industry 4.0 concepts; some of the practices are outlined in the study.

Studies such as Szalavetz (2017) demonstrate the beneficial impact of advanced manufacturing technologies on firm's environmental performance, drawing on interviews conducted with 16 Hungarian manufacturing subsidiaries. Tortorella and Fettermann (2018) used data from a survey carried out with 110 Brazilian companies to examine the relationship between lean production practices and the implementation of Industry 4.0. As for the modelling methodology, Basios and Loucopoulos (2017) propose an approach referred to as Six Sigma DMAIC Enhanced with Capability Modelling approach, for which requirements can be considered from an operational and strategic perspective.

Furthermore, we analyze how studies have used different methods. Figure 5 shows the relations between categories of studies and applied methodologies. It is clear from our SLR that the most common methodology used in the papers under study is LR/SLR or conceptual method.

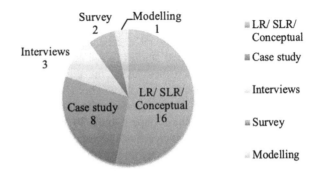

Fig. 4 Number of papers based on the method of research

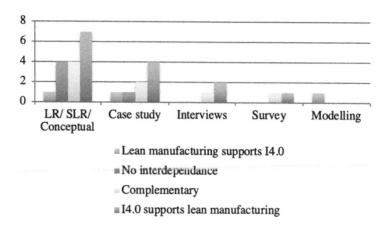

Fig. 5 Categories of papers and applied methodologies

8 Conclusion

This chapter links together literature on knowledge creation (Nonaka, 1994) and task coordination (Thompson, 1967) to examine how the new knowledge stemming from Industry 4.0 and operational excellence is coordinated. According to the literature, new knowledge is created as a result of the recursive interaction of tacit and explicit knowledge that moves through the steps of socialization, externalization, combination, and internalization. When the recursive interaction of tacit and explicit knowledge involves multiple sources of knowledge such as Industry 4.0 and operational excellence, it is also important to take into consideration how these different sources of knowledge are coordinated. In particular, based on Thompson (1967), there are three possible ways of coordinating Industry 4.0 and operational excellence. First, they may be implemented separately. Second, Industry 4.0 and operational excellence may be implemented sequentially and the main issue is to understand whether operational excellence should be implemented before or after new technologies. Third, operational excellence and Industry 4.0 may be implemented together. Literature shows that the way activities are coordinated has profound implications on firm's performance (Barki & Pinsonneault, 2005; Galeazzo et al., 2014) and can greatly impact the way these activities are managed as well as their potential success. To investigate how operational excellence and Industry 4.0 are coordinated, we collect information on past studies focusing on both Industry 4.0 and operational excellence and approach them using the theoretical lens of knowledge creation (Nonaka, 1994) and task coordination (Thompson, 1967).

Our findings draw on a systematic literature review on 374 papers. After an accurate screening of these papers, only 30 were identified as relevant. We can assume that very few studies have focused on both operational excellence and Industry 4.0 literature, thus suggesting a lack of knowledge on this topic. Almost one-third of the papers under study are conference papers; we believe this is due to the emergence and increasing demand of Industry 4.0 topic in academia. Also as shown in Fig. 2 we observe increasing number of publications throughout the past few years, which shows that the interplay between operational excellence and Industry 4.0 is an emerging area. Majority of the papers under study are LR/SLR and conceptual; this finding calls for more empirical research in order to investigate different types of relationship between operational excellence and Industry 4.0.

The main finding of the present SLR shows that operational excellence and Industry 4.0 can be coordinated in different ways. Based on Thompson (1967)'s categorization of task coordination, we identified four categories. The first category includes papers arguing that Industry 4.0 supports operational excellence, implying Industry 4.0 technologies enable operational excellence. The second category includes papers arguing that operational excellence supports Industry 4.0, implying operational excellence enables the implementation of Industry 4.0 technologies. Both these categories suggest that Industry 4.0 and operational excellence should be coordinated sequentially because, as Thompson (1967) highlights, tasks have a sequential interdependence, representing the situation in which each part's outputs are the inputs of another part, and similarly, the inputs that one part uses are the outputs from another part. This type of interdependence requires that parts perform activities in a specified sequence, implying there is an asymmetric exchange between parts. The third category includes papers arguing that Industry 4.0 and operational excellence complement each other, thus implying their coordination occurs simultaneously. This is in line with Thompson (1967) maintaining that the simultaneous coordination is associated with reciprocal interdependence between tasks. Reciprocal interdependence represents a bidirectional exchange of inputs and outputs between parts, meaning that the activities performed by each part pose "contingency for the other" (p. 55). The fourth category includes papers arguing that, though their relevance is well acknowledged, there is no interdependence between Industry 4.0 and operational excellence. This category is closer to pooled interdependence. Pooled interdependence is defined as a situation in which there is absence of workflow between parts. Each part acquires independent inputs and produces independent outputs that contribute to the whole organization. This implies that "each part renders a discrete contribution to the whole and each is supported by the whole" (p. 54). In this case, each part performs activities separately and in any order, without any exchange between parts. Among the four categories, our literature review shows that most of the papers are in the first category, i.e., Industry 4.0 supports operational excellence, suggesting there is a sequential interdependence between operational excellence and Industry 4.0. However, based on these findings, we cannot rule out that the other three categories are not appropriate to describe the way knowledge created by operational excellence and Industry 4.0 is coordinated because of the few papers in our sample.

Finally, our findings highlight that not all Industry 4.0 technologies would support all operational excellence practices and vice versa. For example, Huang et al. (2012) and Zhang et al. (2011) view RFID technology as enabler in providing just-in-time information rendering and real-time traceability. In a study by Brintrup et al. (2010) RFID technology is viewed as a vehicle to achieve operational excellence through automated data collection, assurance of data dependencies, and improvements in production and inventory visibility. Lugert et al. (2018) state that users appreciate a combination of operational excellence-related practices and solutions of Industry 4.0 and that experts request further development of the value stream mapping (VSM); the paper provides a current evaluation of the VSM from an exploratory perspective. Hyun Park et al. (2017) state that artificial intelligence (AI) supports the development and production of high-quality products and that AI technology based on big data is expected to promote the widespread use of quality management.

Future studies should further investigate the relationship between operational excellence and Industry 4.0 in future studies to provide support to our main findings. Another future avenue of research is to investigate the circumstances in which these interdependences work.

References

Arora, A., & Gambardella, A. (1994). The changing technology of technological change: General and abstract knowledge and the division of innovative labour. *Research Policy, 23*(5), 523–532.

Arumugam, V., Antony, J., & Kumar, M. (2013). Linking learning and knowledge creation to project success in Six Sigma projects: An empirical investigation. *International Journal of Production Economics, 141*(1), 388–402.

Barki, H., & Pinsonneault, A. (2005). A model of organizational integration, implementation effort, and performance. *Organization Science, 16*(2), 165–179.

Basios, A., & Loucopoulos, P. (2017) Six sigma DMAIC enhanced with capability modelling. In *2017 IEEE 19th Conference on Business Informatics (CBI)*, vol 2. IEEE, pp. 55–62.

Baumler, J. V. (1971). Defined criteria of performance in organizational control. *Administrative Science Quarterly, 16*(3), 340.

Beard-Gunter, A., Ellis, D. G., & Found, P. A. (2019). TQM, games design and the implications of integration in Industry 4.0 systems. *International Journal of Quality and Service Sciences, 11*(2), 235–247.

Bertoncel, T., Erenda, I., Bach, M. P., Roblek, V., & Meško, M. (2018). A managerial early warning system at a smart factory: An intuitive decision-making perspective. *Systems Research and Behavioral Science, 35*(4), 406–416.

Brintrup, A., Ranasinghe, D., & McFarlane, D. (2010). RFID opportunity analysis for leaner manufacturing. *International Journal of Production Research, 48*(9), 2745–2764.

Buer, S. V., Strandhagen, J. O., & Chan, F. T. (2018). The link between Industry 4.0 and lean manufacturing: Mapping current research and establishing a research agenda. *International Journal of Production Research, 56*(8), 2924–2940.

Byosiere, P., & Luethge, D. J. (2004). Realizing vision through envisioning reality: Strategic leadership in building knowledge spheres. In R. J. Burke & C. Cooper (Eds.), *Leading in turbulent times. Managing in the new world of work* (pp. 243–258). Malden, MA: Blackwell.

Chang, J. J., Hung, K. P., & Lin, M. J. J. (2014). Knowledge creation and new product performance: The role of creativity. *R&D Management, 44*(2), 107–123.

Chang, Y. H., & Yeh, Y. J. Y. (2018). Industry 4.0 and the need for talent: A multiple case study of Taiwan's companies. *International Journal of Product Development, 22*(4), 314–332.

Choo, C. W., & Bontis, N. (Eds.). (2002). *The strategic management of intellectual capital and organizational knowledge*. Oxford: Oxford University Press.

Choo, A. S., Linderman, K., & Schroeder, R. G. (2007). Method and psychological effects on learning behaviors and knowledge creation in quality improvement projects. *Management Science, 53*(2), 437–450.

Colurcio, M. (2009). TQM: A knowledge enabler? *The TQM Journal, 21*(3), 236–248.

Dallasega, P. (2018). Industry 4.0 fostering construction supply chain management: Lessons learned from engineer-to-order suppliers. *IEEE Engineering Management Review, 46*(3), 49–55.

Demartini, M., & Tonelli, F. (2018). Quality management in the industry 4.0 era. In *Proceedings of the Summer School Francesco Turco*, pp. 8–14.

Demeter, K., & Matyusz, Z. (2011). The impact of lean practices on inventory turnover. *International Journal of Production Economics, 133*(1), 154–163.

Deming, W. E. (1994). *The new economics for industry, education, government*. Cambridge,MA: MIT Center for Advanced Engineering Study.

Dierkes, M., Antal, A. B., Child, J., & Nonaka, I. (Eds.). (2003). *Handbook of organizational learning and knowledge*. New York, NY: Oxford University Press.

Droege, S. B., & Hoobler, J. M. (2003). Employee turnover and tacit knowledge diffusion: A network perspective. *Journal of Managerial Issues, 15*, 50–64.

Easterby-Smith, M., & Lyles, M. A. (2003). Introduction: Watersheds of organizational learning and knowledge management. In M. Easterby-Smith & M. A. Lyles (Eds.), *The Blackwell handbook of organizational learning and knowledge management* (pp. 1–15). Malden, MA: Blackwell.

Eleftheriadis, R. J., & Myklebust, O. (2016). A guideline of quality steps towards zero defect manufacturing in industry. In *Proceedings of the International Conference on Industrial Engineering and Operations Management*, pp. 332–340.

Esterhuizen, D., Schutte, C. S., & Du Toit, A. S. A. (2012). Knowledge creation processes as critical enablers for innovation. *International Journal of Information Management, 32*(4), 354–364.

Feldman, M. S., & Pentland, B. T. (2003). Reconceptualizing organizational routines as a source of flexibility and change. *Administrative Science Quarterly, 48*(1), 94–118.

Fine, C. H. (1986). Quality improvement and learning in productive systems. *Management Science, 32*(10), 1301–1315.

Foidl, H., & Felderer, M. (2015, November). Research challenges of industry 4.0 for quality management. In *International Conference on Enterprise Resource Planning Systems* (pp. 121–137). Cham: Springer.

Freedman, B. (2017). The opportunities and challenges of the industrial internet of things. *Quality*, 16VS.

Furlan, A., Dal Pont, G., & Vinelli, A. (2011). Complementarity and lean manufacturing bundles. An empirical analysis. *International Journal of Production and Operations Management, 31*(8), 835–850.

Furlan, A., Galeazzo, A., & Paggiaro, A. (2019). Organizational and perceived learning in the workplace: A multilevel perspective on employees' problem solving. *Organization Science, 30*(2), 280–297.

Furlan, A., & Vinelli, A. (2018). Unpacking the coexistence between improvement and innovation in world-class manufacturing: A dynamic capability approach. *Technological Forecasting and Social Change, 133*, 168–178.

Galbraith, J. R. (1977). *Organization design*. Boston, MA: Addison Wesley.

Galeazzo, A., & Furlan, A. (2018). Lean bundles and configurations: A fsQCA approach. *International Journal of Operations & Production Management, 38*, 513–533.

Galeazzo, A., & Furlan, A. (2019). Good problem solvers? Leveraging knowledge sharing mechanisms and management support. *Journal of Knowledge Management, 23*(6), 1017–1038.

Galeazzo, A., Furlan, A., & Vinelli, A. (2014). Lean and green in action: Interdependencies and performance of pollution prevention projects. *Journal of Cleaner Production, 85*, 191–200.

Galeazzo, A., Furlan, A., & Vinelli, A. (2017). The organizational infrastructure of continuous improvement – An empirical analysis. *Operations Management Research, 10*(1–2), 33–46.

Gilson, L. L., & Goldberg, C. B. (2015). *Editors' comment: So, what is a conceptual paper?* (pp. 127–130).

Gorecky, D., Schmitt, M., Loskyll, M., & Zühlke, D. (2014). Human-machine-interaction in the industry 4.0 era. In *2014 12th IEEE International Conference on Industrial Informatics (INDIN)*, pp. 289–294. IEEE.

Gourlay, S. (2006). Conceptualizing knowledge creation: A critique of Nonaka's theory. *Journal of Management Studies, 43*(7), 1415–1436.

Gully, S. M., Incalcaterra, K. A., Joshi, A., & Beaubien, J. M. (2002). A meta-analysis of team-efficacy, potency, and performance: Interdependence and level of analysis as moderators of observed relationships. *Journal of Applied Psychology, 87*(5), 819.

Hannola, L., Richter, A., Richter, S., & Stocker, A. (2018). Empowering production workers with digitally facilitated knowledge processes–a conceptual framework. *International Journal of Production Research, 56*(14), 4729–4743.

Hickson, D. J., Pugh, D. S., & Pheysey, D. C. (1969). Operations technology and organization structure: An empirical reappraisal. *Administrative Science Quarterly*, 378–397.

Holmström, J., Holweg, M., Khajavi, S. H., & Partanen, J. (2016). The direct digital manufacturing (r)evolution: Definition of a research agenda. *Operations Management Research, 9*, 1–10.

Holsapple, C. W., & Joshi, K. D. (2000). An investigation of factors that influence the management of knowledge in organizations. *The Journal of Strategic Information Systems, 9*(2–3), 235–261.

Huang, G. Q., Qu, T., Zhang, Y., & Yang, H. D. (2012). RFID-enabled product-service system for automotive part and accessory manufacturing alliances. *International Journal of Production Research, 50*(14), 3821–3840.

Hyun Park, S., Seon Shin, W., Hyun Park, Y., & Lee, Y. (2017). Building a new culture for quality management in the era of the Fourth Industrial Revolution. *Total Quality Management and Business Excellence, 28*(9–10), 934–945.

Jiang, X., & Li, Y. (2009). An empirical investigation of knowledge management and innovative performance: The case of alliances. *Research Policy, 38*(2), 358–368.

Johnson, D. W., & Johnson, R. T. (1989). *Cooperation and competition: Theory and research*. Edina, MN: Interaction Book Company.

Khanchanapong, T., Prajogo, D., Sohal, A. S., Cooper, B. K., Yeung, A. C. L., & Cheng, T. C. E. (2014). The unique and complementary effects of manufacturing technologies and lean practices on manufacturing operational performance. *International Journal of Production Economics, 153*, 191–203.

Krishnan, R., Martin, X., & Noorderhaven, N. G. (2006). When does trust matter to alliance performance? *Academy of Management Journal, 49*(5), 894–917.

Lee, J., Bagheri, B., & Kao, H. A. (2015). A cyber-physical systems architecture for industry 4.0-based manufacturing systems. *Manufacturing Letters, 3*, 18–23.

Leyh, C., Martin, S., & Schäffer, T. (2017). Analyzing industry 4.0 models with focus on lean production aspects. In Information technology for management. *Ongoing Research and Development* (pp. 114–130). Springer, Cham

Li, Y. H., Huang, J. W., & Tsai, M. T. (2009). Entrepreneurial orientation and firm performance: The role of knowledge creation process. *Industrial Marketing Management, 38*(4), 440–449.

Linderman, K., Schroeder, R. G., & Sanders, J. (2010). A knowledge framework underlying process management. *Decision Sciences, 41*(4), 689–719.

Linderman, K., Schroeder, R. G., Zaheer, S., Liedtke, C., & Choo, A. S. (2004). Integrating quality management practices with knowledge creation processes. *Journal of Operations Management, 22*(6), 589–607.

Liu, S., Leat, M., Moizer, J., Megicks, P., & Kasturiratne, D. (2013). A decision-focused knowledge management framework to support collaborative decision making for lean supply chain management. *International Journal of Production Research, 51*(7), 2123–2137.

Lugert, A., Batz, A., & Winkler, H. (2018). Empirical assessment of the future adequacy of value stream mapping in manufacturing industries. *Journal of Manufacturing Technology Management, 29*(5), 886–906.

Makhanya, B. B., Nel, H., & Pretorius, J. H. C. (2018). Benchmarking quality management maturity in industry. In *2018 IEEE International Conference on Industrial Engineering and Engineering Management (IEEM)*, EEE, pp 575–579.

Martinez, F. (2019). Process excellence the key for digitalisation. *Business Process Management Journal, 25, 1716*.

Melnyk, S. A., Flynn, B. B., & Awaysheh, A. (2018). The best of times and the worst of times: Empirical operations and supply chain management research. *International Journal of Production Research, 56*(1–2), 164–192.

Merx-Chermin, M., & Nijhof, W. J. (2005). Factors influencing knowledge creation and innovation in an organisation. *Journal of European Industrial Training, 29*(2), 135–147.

Moeuf, A., Pellerin, R., Lamouri, S., Tamayo-Giraldo, S., & Barbaray, R. (2018). The industrial management of SMEs in the era of Industry 4.0. *International Journal of Production Research, 56*(3), 1118–1136.

Moreno-Luzón, M. D., & Begoña Lloria, M. (2008). The role of non-structural and informal mechanisms of integration and coordination as forces in knowledge creation. *British Journal of Management, 19*(3), 250–276.

Nonaka, I. (1994). A dynamic theory of organizational knowledge creation. *Organization Science, 5*(1), 14–37.

Nonaka, I., & Konno, N. (1998). The concept of "Ba": Building a foundation for knowledge creation. *California Management Review, 40*(3), 40–54.

Nonaka, I., & Takeuchi, H. (1995). *The knowledge-creating company: How Japanese companies create the dynamics of innovation*. New York: Oxford University Press.

Nonaka, I., Toyama, R., & Nagata, A. (2000). A firm as a knowledge-creating entity: A new perspective on the theory of the firm. *Industrial and Corporate Change, 9*(1), 1–20.

Nonaka, I., Umemoto, K., & Senoo, D. (1996). From information processing to knowledge creation: A paradigm shift in business management. *Technology in Society, 18*(2), 203–218.

Porter, M. E., & Heppelmann, J. E. (2017). Why every organization needs an augmented reality strategy. *Harvard Business Review, 95*(6), 46–57.

Proudlove, N. C., Vadera, S., & Kobbacy, K. A. H. (1998). Intelligent management systems in operations: A review. *Journal of the Operational Research Society, 49*(7), 682–699.

Ren, S., Zhang, Y., Liu, Y., Sakao, T., Huisingh, D., & Almeida, C. M. V. B. (2019). A comprehensive review of big data analytics throughout product lifecycle to support sustainable smart manufacturing: A framework, challenges and future research directions. *Journal of Cleaner Production, 210*, 1343–1365.

Roblek, V., Meško, M., & Krapež, A. (2016). A complex view of Industry 4.0. *SAGE Open, 6*(2), 1–11.

Rossini, M., Costa, F., Tortorella, G. L., & Portioli-Staudacher, A. (2019). The interrelation between Industry 4.0 and lean production: An empirical study on European manufacturers. *The International Journal of Advanced Manufacturing Technology, 102*(9–12), 3963–3976.

Roy, D., Mittag, P., & Baumeister, M. (2015). Industrie 4.0 – Einfluss der Digitalisierung auf die fünf Lean-Prinzipien Schlank vs. Intelligent. *Product Management, 20*, 27–30.

Rüttimann, B. G., & Stöckli, M. T. (2016). Lean and Industry 4.0 – Twins, partners, or contenders? A due clarification regarding the supposed clash of two production systems. *Journal of Service Science and Management, 9*, 485–500.

Sanders, A., Elangeswaran, C., & Wulfsberg, J. P. (2016). Industry 4.0 implies lean manufacturing: Research activities in industry 4.0 function as enablers for lean manufacturing. *Journal of Industrial Engineering and Management, 9*(3), 811–833.

Santos, F. M. (2003). The role of information technologies for knowledge management in firms. *International Journal of Technology, Policy and Management, 3*(2), 194–203.

Schroeder, R. G., Linderman, K., Liedtke, C., & Choo, A. S. (2008). Six Sigma: Definition and underlying theory. *Journal of Operations Management, 26*(4), 536–554.

Schulze, H., & Bals, L. (2018). Implementing sustainable supply chain management: A literature review on required purchasing and supply management competences. In *Social and Environmental Dimensions of Organizations and Supply Chains* (pp. 171–194). Cham: Springer.

Shea, G. P., & Guzzo, R. A. (1987). Groups as human resources. *Research in Personnel and Human Resources Management, 5*, 323–356.

Sin, A. B., Zailani, S., Iranmanesh, M., & Ramayah, T. (2015). Structural equation modelling on knowledge creation in Six Sigma DMAIC project and its impact on organizational performance. *International Journal of Production Economics, 168*, 105–117.

Subramaniam, M., & Youndt, M. A. (2005). The influence of intellectual capital on the types of innovative capabilities. *Academy of Management Journal, 48*(3), 450–463.

Szalavetz, A. (2017). The environmental impact of advanced manufacturing technologies: Examples from Hungary. *Central European Business Review, 6*(2), 18–29.

Thompson, J. D. (1967). *Organizations in action*. New York: McGraw-Hill.

Tortorella, G. L., de Castro Fettermann, D., Frank, A., & Marodin, G. (2018). Lean manufacturing implementation: Leadership styles and contextual variables. *International Journal of Operations & Production Management*.

Tortorella, G. L., & Fettermann, D. (2018). Implementation of Industry 4.0 and lean production in Brazilian manufacturing companies. *International Journal of Production Research, 56*(8), 2975–2987.

Tranfield, D., Denyer, D., & Smart, P. (2003). Towards a methodology for developing evidence-informed management knowledge by means of systematic review. *British Journal of Management, 14*(3), 207–222.

Trotta, D., & Garengo, P. (2018). Industry 4.0 key research topics: A bibliometric review. In *2018 7th International Conference on Industrial Technology and Management (ICITM) IEEE* (pp. 113–117).

Tsai, M. T., & Li, Y. H. (2007). Knowledge creation process in new venture strategy and performance. *Journal of Business Research, 60*(4), 371–381.

Urbinati, A., Bogers, M., Chiesa, V., & Frattini, F. (2019). Creating and capturing value from Big Data: A multiple-case study analysis of provider companies. *Technovation, 84*, 21–36.

Vaccaro, A., Veloso, F., & Brusoni, S. (2009). The impact of virtual technologies on knowledge-based processes: An empirical study. *Research Policy, 38*(8), 1278–1287.

Van de Ven, A. H., Delbecq, A. L., & Koenig Jr, R. (1976). Determinants of coordination modes within organizations. *American Sociological Review*, 322–338.

Van Der Vegt, G., Emans, B., & Van De Vliert, E. (2000). Team members' affective responses to patterns of intragroup interdependence and job complexity. *Journal of Management, 26*(4), 633–655.

Van der Vegt, G. S., Emans, B. J., & Van De Vliert, E. (2001). Patterns of interdependence in work teams: A two-level investigation of the relations with job and team satisfaction. *Personnel Psychology, 54*(1), 51–69.

Vogelsang, K., Packmohr, S., Liere-Netheler, K., & Hoppe, U. (2018, September). Understanding the transformation towards industry 4.0. In *International Conference on Business Informatics Research* (pp. 99–112). Cham: Springer.

Yin, Y., Stecke, K. E., & Li, D. (2018). The evolution of production systems from Industry 2.0 through Industry 4.0. *International Journal of Production Research, 56*(1–2), 848–861.

Zahra, S. A., Ireland, R. D., & Hitt, M. A. (2000). International expansion by new venture firms: International diversity, mode of market entry, technological learning, and performance. *Academy of Management Journal, 43*(5), 925–950.

Zhang, J., Ong, S. K., & Nee, A. Y. C. (2011). RFID-assisted assembly guidance system in an augmented reality environment. *International Journal of Production Research, 49*(13), 3919–3938.

Achieving Circular Economy Via the Adoption of Industry 4.0 Technologies: A Knowledge Management Perspective

Valentina De Marchi and Eleonora Di Maria

Abstract The chapter discusses the relationship between knowledge management, circular economy strategies and investments in Industry 4.0 technologies. Based on an empirical analysis of about 200 Italian manufacturing firms adopting such technologies, the chapter maps the knowledge implications of adoption comparing firms that achieved and did not achieved sustainability results, with a special focus on circular economy. Results highlight that environmental sustainability is often an unexpected, non-planned result of the adoption, and that sustainability outcomes entail more often the adoption of robotics and augmented reality technologies. Interestingly, adoption resulting in sustainability outcomes is more likely when the technology is adopted in the production process. Stronger intra-firm collaboration and engagement with customers is also more likely to take place for green adopter. Few differences emerge if comparing circular with eco-efficiency-oriented outcomes.

1 Introduction

Firms are increasingly challenged to transform their activities and products in order to ensure they are taking responsibility of the environmental impacts. There is growing attention on how digital technologies and specifically technologies related to the so-called Industry 4.0 may impact on the environmental sustainability of firms and on the opportunity to achieve goals related to the circular economy framework (Beatriz et al., 2018). More specifically, those new technologies can expect to support the implementation of circular economy (CE), conceived as a *"a regenerative system in which resource input and waste, emission, and energy leakage are minimised by slowing, closing, and narrowing material and energy loops. This can be achieved through long-lasting design, maintenance, repair, reuse,*

V. De Marchi (✉) · E. Di Maria
Department of Economics and Management 'Marco Fanno', University of Padova, Padova, Italy
e-mail: valentina.demarchi@unipd.it

remanufacturing, refurbishing, and recycling" (Geissdoerfer, Savaget, Bocken, & Hultink, 2017, p. 759). But is this connection feasible?

Industry 4.0 technologies shape manufacturing processes and push forward the possibility to develop smart factories and smart products, enhancing efficiency and better control within the factory and the entire value chain (Roblek, Meško, & Krapež, 2016). Recent studies (Cezarino, Liboni, Oliveira Stefanelli, Oliveira, & Stocco, 2019) highlight the interdependence between digitalization and circularity within the firm and the value chains. According to the review of those authors, Industry 4.0 technologies sustain both efficiency and flexibility, allowing also the firm to have better resource allocation and control over the processes both within and outside its boundaries. Prior studies have suggested that Industry 4.0 technologies are valuable tools for the implementation of circular economy in terms of new business models (Despeisse et al., 2017), new product development, and customer involvement and value chain reconfiguration (Hazen, Mollenkopf, & Wang, 2017; Unruh, 2015). In this view, there are also the premises for the firm to increase its knowledge on how resources are used and which are the dynamics of flows of resources and products both upstream and downstream (Tseng, Tan, Chiu, Chien, & Kuo, 2018).

In this chapter, we aim at further exploring the link between 4.0 technologies and circularity, discussing if and to what extent Industry 4.0 technologies can enhance knowledge acquisition as well as support the firm in better managing such knowledge for sustainability purposes. This goal is particularly useful as within the circular economy discussion a remarkable dimension is represented by how to possibly measure the "circularity" at the different levels and to develop a shared set of indicators useful to capture the implementation of circular economy, its impacts on the environment, and the related economic dimensions (Moraga et al., 2019; Saidani, Yannou, Leroy, Cluzel, & Kendall, 2019). In this perspective, Industry 4.0 technologies may sustain this knowledge creation and transfer through a set of tools (i.e., sensors, IoT) and processes (i.e., additive manufacturing). However, little theoretical and empirical evidence has been developed so far on this issue through the lenses of knowledge management.

Based on an empirical analysis of about 200 Italian adopting firms in selected manufacturing industries, in this chapter we investigate the relationship between circular economy strategies of firms and the investments in Industry 4.0 solutions from a knowledge management point of view. In particular, the chapter analyzes the implications of knowledge management for the circular economy framework and the opportunities and challenges related to Industry 4.0 technologies in this scenario.

2 Circular Economy in a Knowledge Management Framework

2.1 *Environmental Innovation and Knowledge Sourcing*

In order to take into account the environmental consequences of their production, firms introduce sustainability-oriented innovations, by focusing on a better or diverse input selection and use or by investing in order to reduce product impacts at the end of its life (i.e., through product modularity) (Jay & Gerard, 2015). A vast literature has studied these innovations, highlighting that they are peculiar in the (positive environmental) externalities they can produce but also, and possibly most interestingly, in the way the new products and processes are developed (Bansal & Grewatsch, 2019; Cainelli, De Marchi, & Grandinetti, 2015).

In particular, literature on environmental innovations highlights the need for firms to develop collaborations with external partners in order to effectively introduce green innovations. Starting from the knowledge management perspectives, researchers pointed out that a single firm, alone, is not able or might not find convenient to develop or exploit all the required resources to transform its products and processes toward sustainability (De Marchi & Grandinetti, 2013). Interestingly, cooperation for innovation has been found to be more relevant for R&D environmental-related innovation than for non-environmental ones. Pittaway, Robertson, Munir, Denyer, and Neely (2004) highlight that through cooperation firms benefit in terms of innovation capacity. Such strategic approach is also useful in case of firms with strong internal innovation capacities—i.e., large R&D investments—since the firm exploits connections with other partners—from suppliers to universities or other KIBS—to speed up knowledge acquisition or reduce costs and risks.

According to the literature, several actors can be supportive of eco-innovation at firms (Cainelli et al., 2015), supporting innovation trajectories, and sustain firm's capabilities related to green goals. As far as upstream partners are concerned, specialized suppliers might provide key knowledge to the firm in terms of selected material or components, in addition to positive green practices such as just-in-time solutions and the like (De Marchi & Di Maria, 2019). On the other hand, they are also the sources of inputs and components, stressing the interdependence among actors within the value chain to achieve more extended sustainability results (Seuring & Müller, 2008).

In addition to upstream knowledge transfer, also downstream partners can become important sources of knowledge concerning product as well as process development (Cervellon & Wernerfelt, 2012; Darnall, Jolley, & Handfield, 2008; Mylan, Geels, Gee, McMeekin, & Foster, 2015) both if we consider final consumers but also, and possibly more interestingly, if we take into account retailers or other business-to-business customers. As far as the final consumers are concerned, customers may provide the firm with knowledge inputs in terms of product durability, context of use, and performance, signaling eventually problems, within the increasing shift in the demand side toward greener consumption habits. At the same time, customers can ask for knowledge from the firm in terms of traceability of the products and information on their sourcing (Lacy & Rutqvist, 2016). This may depend on their difficulties in evaluating the level of sustainability of products and the potential advantages related to their potentially green buying behaviors (Devinney, Auger, & Eckhardt, 2010). Also the role of buyers in pushing sustainability is widely described in the literature (De Marchi, Di Maria, Krishnan, & Ponte, 2019; De Marchi, Di Maria, & Ponte, 2013). Those business actors may transfer to the firm knowledge concerning market requests as well as further support suppliers—as discussed above—in their sustainability sourcing processes.

2.2 Collaborations for the Circular Economy

Within the broad debate related to sustainability, recent contributions refer to CE as a new economic paradigm, which substitutes the linear mechanism of "make, use, dispose" with a closed-loop approach in the use of resources. According to this emerging framework, the entire value chain of the firm – from suppliers to final customers—as well as the whole business (and natural) ecosystem can benefit from an integrated approach on the production, selection, and use of resources as inputs as well as outputs (products) (Geissdoerfer et al., 2017; Ghisellini, Cialani, & Ulgiati, 2016; Webster & MacArthur, 2017). CE challenges innovation processes of environmental-oriented firms in the direction of lower use of resources or waste-based inputs. In this direction, it extends the supply base beyond the established suppliers, to also involve new ones based on potential new inputs coming from waste or reuse practices (Lacy & Rutqvist, 2016; Webster & MacArthur, 2017). Moreover, CE emphasizes efficient manufacturing processes in terms of leakage avoidance but also shortening supply chains (i.e., distributed manufacturing processes).

Recent studies suggest that CE can be interpreted both as "a holistic concept and operational tool" (de Jesus, Antunes, Santos, & Mendonça, 2019, p. 1494). In this sense, CE can be understood as a transformative approach toward eco-innovation, which combines both hard (i.e., R&D-driven products) and soft (i.e., business models) types of knowledge. In other words, CE entails the introduction of innovations, which can be considered as a subgroup of eco-innovations (Horbach & Rammer, 2019).

According to Geissdoerfer et al. (2017) the CE concept relies on multiple theoretical schools and with respect to sustainability CE emerges as a framework strongly oriented to the better use of resources, reducing waste and environmental leakages. In this respect, it generates more complexity due to its intrinsic systemic nature. Among the five fundamental traits of CE identified by the Ellen MacArthur Foundation, we can stress the need "to think in systems" suggesting the request of exploring and understanding the rich set of interdependencies existing among multiple layers of the manufacturing activity and beyond, since "components are considered in connection to their economic, operational, environmental, and social impacts and contexts" (Esposito, Tse, & Soufani, 2018, p. 9). Beyond the material management issue (from inputs, to manufacturing up to waste management)—tightly coupled with innovation processes—CE also includes innovation in terms of value creation through new business models connected to emerging CE strategies (Hopkinson, Zils, Hawkins, & Roper, 2018; Lacy & Rutqvist, 2016; Ünal & Shao, 2018). Cainelli et al. (2015) argue that complexity and the systemic dimension are among the "causes" of a specific trait of eco-innovations, which drives the higher need for cooperation with external partners. Considering that CE innovations have those characteristics that create differences with respect to traditional but also of eco-efficient-related innovations, we can infer that the firm implementing CE innovations needs to rely on external sources of knowledge.

This argument is supported by recent studies. Brown, Bocken, and Balkenende (2019) clearly map the drivers and barriers for collaboration within the circular-oriented innovation. According to those authors, the challenges the new emerging CE framework is posing – related to a new systemic approach that includes product design, circular manufacturing processes, as well as CE recovery strategies (Bocken, de Pauw, Bakker, & van der Grinten, 2016)—ask for new knowledge management strategies. In fact, circular-oriented firms (and innovators) have to expand their sources of knowledge and networks to cope with such a technical, economic, and social complexity.

3 Circular Economy and the Opportunities Related to Industry 4.0

The development and implementation of CE strategies have to take into account the emerging new digital technological scenario. There is a growing body of literature from both the sustainability perspective and the digital perspective that suggests the need to exploit the convergence between CE and industry 4.0 technologies (Beatriz et al., 2018; Rajput & Singh, 2019), as such technologies have the potential to enable higher circular economy results. Industry 4.0 technologies open interesting opportunities to sustain environmental sustainability taking into account inputs, process, and product management within the value chains (Stock & Seliger, 2016), considering the need of lower use of resources as inputs, reduced waste generation, or pollution consequences during production or consumption activities (Chen et al., 2015; Yeo, Pepin, & Yang, 2017). Such a link has been postulated considering different types of technologies.

Recent research published in a special issue on *California Management Review* explores (mainly theoretically) the process of leveraging 3D printing for circularity. 3D printing allows better use of resources within the firm, producing what is needed where and when it is needed (Despeisse et al., 2017; Unruh, 2018). From this point of view, customers become sources of knowledge to provide the firms with inputs on the products they desire as well as on the product requirements to be developed. They may also become active parts in the manufacturing processes themselves. This trend is linked to the rise of new types of customers that are also makers, that is they exploit digital technologies and specifically 3D printing to invest in product manufacturing at home (Anderson, 2012). Firms can leverage on such customers' skills and competence to include them into circular processes (Bressanelli, Adrodegari, Perona, & Saccani, 2018).

Besides 3D printing, new technological solutions related to big data, artificial intelligence, and IoT solutions for improved and extended information gathering and management increase firm's control over internal as well as external processes and relationships with actors of the value chain and consumers (Ellen MacArthur Foundation, 2019; Huberty, 2015; Rajput & Singh, 2019). From this perspective, smart, connected products support knowledge gathering from customers (Porter & Heppelmann, 2015), and they offer the opportunity to enhance product lifecycle and augment value embedded into the product through additional services (product-as-a-service) (Coreynen, Matthyssens, & Van Bockhaven, 2017).

More broadly, Industry 4.0 technologies related to cyber-physical systems, IoT, and cloud computing can generate benefits from a circular economy point of view since they allow design for circularity based on the information gathered from customers as well as through the whole production process (de Sousa Jabbour, Jabbour, Foropon, & Filho, 2018). Within the factory, those sets of technologies might drive cleaner production in terms of better control over resource management (machines may become new knowledge sources through remote monitoring, in addition to workers). Additionally, it might support mass production and new customization opportunities. Moreover, they may allow suppliers being included in the loop by exchanging knowledge concerning sourcing performance, quality, and input characteristics.

In a CE framework, where the attention for measuring is high, data-driven strategies may be of particular relevance to support firms in the achievement of greater CE results (Tseng et al., 2018) through a deeper understanding of use of resources and their dynamics within the value chain (across time and space) based on distributed knowledge sources.

While there is increasing attention toward such issues, very few studies have empirically verified if and to what extent such technologies are enablers of CE outcomes. Based on the abovementioned theoretical premises and via data from an original survey, we are interested in investigating if, and under what circumstances, the adoption of industry 4.0 technological solutions can lead to improved environmental outcomes.

4 Empirical Analysis

4.1 Data and Empirical Setting

In order to answer our research questions, we carried out a survey in the period May–December 2017. The survey addressed firms located in the North of Italy. The choice is due to the relevance they have for the Italian gross domestic product (GDP) and for the national competitiveness in the international markets. The universe consisted of 7714 manufacturing firms drawn from AIDA database selected in the industry considered (namely automotive, rubber and plastics, electronic appliances, lightning, furniture, eyewear, jewelry, and sport equipment) (i.e., the Made in Italy sectors) and with a turnover higher than one Million Euro (in industries characterized by the presence of industrial districts,[1] firms with a turnover lower than one Million Euro have been also considered).

Based on a structured questionnaire submitted through CAWI methodology to entrepreneurs, chief operations officers, or managers in charge for manufacturing and technological processes, firms have been contacted and 1229 firms (15.3% of the universe) answered to the survey. The questionnaire assessed the adoption of the following technologies: (1) robotics, (2) additive manufacturing, (3) IoT and intelligent products, (4) big data and cloud, (5) scanner 3D, and (6) augmented reality. Non-adopting firms were asked about reasons for such a choice; adopters were asked about how and why such technologies have been adopted, what the results achieved, and which difficulties encountered.

4.2 Measuring Adoption and Circularity

Besides other types of details related to Industry 4.0 technological investment strategies, the questionnaire collects several useful information to assess the sustainability effects achieved thanks to the adoption of Industry 4.0 technologies. We used this question to build a variable that allows us identifying the so-called CE adopters and to distinguish them both from firms that did not achieve any sustainability results (*non-green*) and from those that achieved sustainability outcomes that are not related to CE but that rather implement an eco-efficient approach (*eco-efficiency*) (Geissdoerfer et al., 2017).

Companies were specifically asked to rate, on a scale from 1 to 5, if the introduction of Industry 4.0 technologies drove any environmental sustainability impact, considering each of the following aspects:

1. Reduction of production waste;
2. Reduction of inputs used (including energy, materials);
3. Reduction of process-related environmental impacts (e.g., on air or water);
4. Adoption of more sustainable inputs (e.g., recycled or recyclable materials);
5. Use of firm's waste in the production process;
6. Use of waste coming from other sectors/firms as inputs.

The variable GREEN ADOPTERS is a dummy taking value 1 if the company reports that the adoption of Industry 4.0 technologies resulted in a strong (4) or very

---

[1]Lightning, eyewear, jewelry, and sport equipment.

strong (5) reduction in any of the 6 elements reported. ECO-EFFICIENCY takes on value 1 if firms reported a strong or very strong impact as for the point from 1 to 3. CIRCULAR equals 1 if companies reported a strong or very strong impact as for any of the elements of the above list ranging from 4 to 6. The two variables do not overlap: in case a firm had reported both eco-efficiency and CE outcomes, we classified it as CE.

In the following, we report results of the multivariate analysis of variance (chi-square test and *t*-tests) to investigate the relationship between Industry 4.0 technologies adoption and the reduction of environmental burden of firms' activity, comparing adopters that reported important sustainability outcomes (GREEN ADOPTERS) with those that did not, distinguishing between firms that achieved CE outcomes from those that achieved just eco-efficiency-related outcomes.

4.3 Sustainability Strategies and Outcomes

A first question addressed regards the fact if the achievement of sustainability-related outcomes of the adoption has been intended or unintended. To this purpose, we use a question asking firms about the motivation of the adoption of 4.0 s, rating, on a scale from 1 (null) to 5 (very much), the importance of 11 items as motivations to invest in 4.0 s, including environmental sustainability.[2] The dummy GREEN DRIVER takes values 1 if the company reported that environmental sustainability was a very (4) or very much (5) relevant motivation for the introduction of 4.0 s. Results are reported in Table 1.

One firm out of four (27.3%) among those part of our sample reported that sustainability concerns have been among the key drivers supporting their investment in Industry 4.0 technologies. This result stresses how such technologies can support explicit environmental sustainability strategies carried out by firms, where digital transformation of business activities and processes is perceived as a means for sustainability. Interestingly, the share of companies that achieved important environmental benefits following Industry 4.0 technologies adoption is almost double (41.7%). While just one-fourth of the companies had a clear sustainability strategy before introducing 4.0 technologies, a much larger share did recognize such an opportunity following their introduction. This is an important result since it suggests that environmental benefits can be considered a sort of "by-product" of the investment in Industry 4.0 technologies, where such technologies enable firms to also gain from an environmental point of view through technological investments motivated by other reasons (the main one is related to better customer service). Moreover, it

---

[2]The full list of items include the following: (1) efficiency of internal processes, (2) increase product variety, (3) opening new market opportunities, (4) maintain production in Italy, (5) reshoring, (6) maintain international competitiveness, (7) imitation of competitors, (8) increase customer service, (9) to respond to request by large buyers, (10) to adapt to an industry standard, and (11) environmental sustainability.

Table 1 Role of 'greening' motivations and outcomes and adoption of Industry 4.0 technologies

		GREENING AS DRIVER		
		No	Yes	Total
GREEN ADOPTERS	No	66	11	77
		50.0%	8.3%	58.3%
	Yes	30	25	55
		22.70%	18.9%	41.7%
	Total	96	36	132
		72.7%	27.3%	100%

Note: Chi-square statistics for statistical significance: Pearson chi^2(1) = 15.71, Pr = 0.000

could push these firms to further pursue green strategies with a more proactive behavior (Bianchi & Noci, 1998).

Table 1 also reports the co-occurrence of GREEN ADOPTERS and GREEN DRIVER: while the share of companies that failed to reach their sustainability goal is quite small (8.3%), 18.9% of the companies realized important environmental benefits that was not initially planned, or, better yet, was not the principal driver for the introduction of Industry 4.0 technologies. The results of the Pearson chi-square test suggest that we can reject hypothesis H0, that is a significant relation between the two dummies considered does not exist.

4.4 Investments in Industry 4.0 Technologies and Sustainability Outcomes

Under the umbrella term Industry 4.0 technologies there is a broad array of technologies, which require different investments levels, are adopted in different value chain activities, and might provide different outcomes. Therefore, we are interested in understanding which of the technologies considered are better supportive of sustainability results and in particular to CE ones. Table 2 reports the number and the share of companies adopting the seven types of Industry 4.0 technologies listed, considering the sustainability impact achieved thanks to the adoption of 4.0 s.

Robotics, big data, and additive manufacturing are by far the most adopted technologies in the sample, but their diffusion across the different groups of firms considered in this chapter is quite different.[3] While those three technologies are the most adopted in all of them, indeed, the relative share of firms that did adopt them varies significantly. As supported by the Pearson chi-square tests, GREEN ADOPTERS show a significantly high degree of adoption of robotics and of augmented reality technologies (63.6% vs. 42.3% and 23.6% vs. 7.7%). Significant

---

[3]Caveat: sample size is small and the adoption of certain technologies such as additive manufacturing is still not so widespread that a cautionary approach to the analysis is needed.

Table 2 Industry 4.0 technologies adoption and sustainability outcomes

	Non-green adopters	Green adopters	Sig.	of which Eco-efficiency	Circular	Sig.
Robotics	42.3%	63.6%	**	64.0%	71.4%	
	33	35		32	20	
Additive manufacturing	39.7%	36.4%		34.0%	46.4%	
	31	20		17	13	
Big data	42.3%	47.3%		46.0%	60.7%	
	33	26		23	17	
3D scanner	24.4%	18.2%		16.0%	32.1%	**
	19	10		8	9	
Augmented reality	7.7%	23.6%	***	22.0%	28.6%	***
	6	13		11	8	
IoT	21.8%	32.7%		32.0%	28.6%	
	17	18		16	8	
Num. of firms	78	55		50	28	
	100%	100%		100%	100%	

Note: The fourth column reports statistically significant differences between NON-GREEN and GREEN ADOPTERS, the seventh between ECO-EFFICIENCY and CIRCULAR, based on Pearson Chi-squared statistics, considering confidence levels: ***$p < 0.01$, **$p < 0.05$

Table 3 Features of 4.0 adoption and sustainability outcomes

	Non-green adopters	Green adopters	Sig.	of which Eco-efficiency	Circular	Sig.
Number of 4.0s adopted (0–6)	1.78	2.22	**	2.68	2.14	
Amount of investment in 4.0s (%, 0–100)	9.64	13.10		14.33	12.85	
Adoption of 4.0s in production (D)	0.66	0.87	***	0.82	0.92	
Adoption of 4.0s in NPD (D)	0.45	0.45		0.54	0.42	
Number companies	78	55		50	28	

Note: The fourth column reports statistically significant differences between NON-GREEN and GREEN ADOPTERS, the seventh between ECO EFFICIENT and CIRCULAR, based on Pearson Chi-squared (between qualitative variables) and t-test (between qualitative and quantitative variables) statistics, considering confidence levels: ***$p < 0.01$, **$p < 0.05$. D means Dummy

differences also emerge when comparing eco-efficient and circular adopter: indeed, circular firms are significantly more likely to adopt 3D scanners and augmented reality technologies. On average, firms that reported environmental benefits are adopting a higher number of technologies (2.22 vs. 1.78 out of 6), a figure driven by the adoption of CIRCULAR firms (2.68, see Table 3). In fact, CIRCULAR firms show the highest share of Industry 4.0 technologies among the types of firms considered, except for IoT which is higher for companies focused just on improving their efficiency and reducing process emissions (ECO-EFFICIENCY).

Interestingly, despite the number of technologies adopted differs significantly across the groups, the amount of money invested to adopt and develop them is not significantly different, if measured as incidence on firm turnover (INVESTMENT IN 4.0S, see Table 3). We might interpret this figure as the fact that the adoption of diverse technologies might allow firms a larger "creativity space," which allow them achieving a wider set of outcomes, including sustainability-related ones.

However, they differ significantly in terms of how those technologies have been implemented in the firm, which can be proxied by the stage of the value chain for which Industry 4.0 technologies have been adopted. Firms that reduced environmental impacts after the introduction of Industry 4.0 technologies are significantly more likely to have adopted them in the production process (87% of the firms in the first group vs. 66% of the second), but not in terms of the development of new products. This result somehow contradicts the expectations in the literature that digitalization of activities might take place especially for innovation purposes. On the contrary, Industry 4.0 technologies seem to allow greater results in terms of environmental sustainability when it comes to production processes and might be strictly related to the relevance of robotics. In this context, no significant differences emerge across eco-efficient firms and CE ones.

4.5 Industry 4.0 and the Knowledge Management Profile of Green Adopters

Considering the specificities of green innovation, even stronger when it comes to innovations related to the circular economy, we expect that the adoption of 4.0 s might have different implications in terms of knowledge management.

Accordingly, we investigate different aspects related to knowledge management, considering both dynamics internal and external to the firm, with a special focus on customers, which are reported in Table 4.

As for the internal dynamics, the adoption of 4.0 technologies does not require differential training for the workforce for its implementation across the groups: this results make sense, as we expect the intensity of such trainings should depend on the (mix of) technologies adopted rather than on the outcomes that can be achieved thanks to their adoption. Interestingly, however, significant differences emerge in terms of the intra-firm collaborations, possibly enhanced with the aim of taking the most out of the Industry 4.0 technologies introduction. Indeed, GREEN ADOPTERS are more likely to have increased collaboration among workers and among firm's functions, which is coherent with the view that green innovation has a higher degree of system dimension. The adoption of Industry 4.0 technologies also allowed the creation of new knowledge: an occurrence that is higher for GREEN

Table 4 Knowledge management implications considering for 4.0 s adoption and sustainability outcomes

	Non-green adopters	Green adopters	Sig.	of which Eco-efficiency	Circular	Sig.
Increased training	2.63	3.17		3.23	3.11	
Increased collaboration...						
between workers and suppliers	1.95	2.41		2.15	2.64	
among workers	1.86	2.72	***	2.69	2.75	
among firm's functions	2.01	3.07	***	2.96	3.18	
Creation of new knowledge...						
to improve processes	2.91	3.70	**	3.77	3.64	
to improve products	2.80	3.33	**	3.27	3.39	
Increased servitization	2.47	2.98	**	2.96	3.00	
Increased traceability	2.14	2.80	**	2.52	3.07	*
Increased cooperation with customers in new product development	2.09	3.04	***	2.52	3.54	
Increased role of customers in production	1.66	2.20		1.52	2.86	***
Num. companies	78	55		50	28	

Note: Variables can take on values from 1 to 5, measuring the level of importance form low to very high. The fourth column reports statistically significant differences between NON-GREEN and GREEN ADOPTERS, the seventh between ECO EFFICIENT and CIRCULAR, based on Pearson Chi-squared (between qualitative variables) and t-test (between qualitative and quantitative variables) statistics, considering confidence levels: $***p < 0.01$, $**p < 0.05$, $*p < 0.1$

ADOPTERS than non-green ones both as it comes to processes and products, which highlights the innovation opportunities that such technologies might entail.

Coherently with the existing literature on environmental innovation interesting differences across the groups emerge, when we consider implications regarding the engagement with the customers. Almost all the variables considered, indeed, point to the development of a closer integration with the customers after the implementation of the technology. This evidence comes in place both as we investigate the direct cooperation with customers (which is significantly different between GREEN and NON-GREEN ADOPTERS just in the case of new product development, not of production) and the increased level of "indirect" interaction that is allowed by the use of the product. Indeed, respondents claimed an increase in the service offered and an enhanced possibility to control the product during its use.

As for other analyses performed in this chapter, the difference between firms that adopt 4.0 s and achieved eco-efficiency rather than circular type of outcomes is not as stark as those characterizing firms that did achieve any of those outcomes (GREEN ADOPTERS) with respect to those that did not (NON-GREEN ADOPTERS). An interesting exception is the variable that captures the importance of the engagement of the customers during the manufacturing processes, which is significantly high for the firms achieving circular outcomes and low for those that did not. Such evidence supports the literature suggesting that the adoption of 4.0 s might allow greater integration along the partners of the value chain and is possibly connected with the possibility to collect waste to be upcycled or recycled.

5 Conclusions

The chapter provides empirical evidence of the connection between Industry 4.0 technologies and CE strategies by also investigating the perspective of knowledge management. Theoretical debate has suggested a promising positive scenario for circular-oriented firms in the adoption of such a set of technologies to control use of resources and monitor internal and external processes. The complexity and systemic dimension of CE should push firms to search for knowledge partners both within and outside the firm's boundaries.

Our analysis suggests a positive relationship between Industry 4.0 technologies and green adopters, with them being focused on CE outcomes or eco-efficiency ones. Differences also emerge in terms of specific technologies adopted and implications in terms of the activities within the value chains those technologies are implemented in. In particular, consistently with studies suggesting a strong role of Industry 4.0 technologies in the production sphere (Stock & Seliger, 2016), results stress how green and non-green adopters differ specifically in the adoption of those technologies (i.e., robotics) in operations processes. This outcome provides evidence of the centrality of manufacturing within the environmental sustainability and CE framework and the enabling role of the technologies part of the so-called fourth industrial revolution. Despite the attention on 3D printing as key enabling technologies for CE, no specific evidence appears in this regard. Further research should investigate if this unexpected result is driven by industry or country specificities or is an element that can be also tracked in other empirical settings.

It is also worth noting that in addition to firms that proactively adopted such technologies to pursue green strategies, there is also a group of adopters that obtain green outcomes from the adoption as an unexpected result. This means that those technologies are not just enablers but also amplifiers—they might further spread the attention toward CE and environmental sustainability since they provide firms with new tools to help control and measure the use of resources (product inputs or energy) and further boost their resource efficiency. From this point of view, Industry 4.0 technologies may become a driver for CE, in addition to other drivers identified in the literature (de Jesus et al., 2019).

From a knowledge management perspective, a clear contrast between green and non-green adopters emerges. In the case of "green adopters," investments and use of Industry 4.0 technologies are strongly related to a stronger emphasis on intra-firm collaboration. Our study suggests that green adopters are more likely to create new knowledge (both at the process and product level) thanks to the adoption of new technologies. A deeper exchange of knowledge with the customers takes place for sustainability outcomes to be achieved—which is especially the case to develop new products. Another interesting difference regards the knowledge implications for the product or service offered. The level of services attached to products (servitization) is more likely to be increased via Industry 4.0 technologies (even if differences in the level of IoT adoption seem not to exist between the two groups of adopting firms), consistently with CE studies (Lacy & Rutqvist, 2016). Additionally, it is more likely for new products to be traced—a result that is coherent with interest to "close the loop," yet that entails also interesting knowledge management implications that should be further studied in depth via case studies.

Our study enriches also the theoretical discussion focused on describing the characteristics of CE with respect to the wider debate on environmental sustainability. At least from the angle of the adoption of Industry 4.0 technologies, analysis described in the chapter suggests that no specific differences emerge in general between eco-efficiency and CE, while there is stronger distinction between firms with or without green outcomes. Further research should investigate this issue more extensively, together with the dynamics of knowledge creation based on collaborative relationships with customers and within the operations processes.

References

Anderson, C. (2012). *Makers: The new industrial revolution*. New York: Crown Business Books.
Bansal, P., & Grewatsch, S. (2019). The unsustainable truth about the stage-gate new product innovation process. *Innovation: Organization and Management*, 1–11. https://doi.org/10.1080/14479338.2019.1684205.
Beatriz, A., De Sousa, L., Charbel, J., Chiappetta, J., Godinho, M., & David, F. (2018). Industry 4.0 and the circular economy: A proposed research agenda and original roadmap for sustainable operations. *Annals of Operations Research, 270*(1), 273–286. https://doi.org/10.1007/s10479-018-2772-8.
Bianchi, R., & Noci, G. (1998). "Greening" SMEs' competitiveness. *Small Business Economics, 11*, 269–281.
Bocken, N. M. P., de Pauw, I., Bakker, C., & van der Grinten, B. (2016). Product design and business model strategies for a circular economy. *Journal of Industrial and Production Engineering, 33*(5), 308–320. https://doi.org/10.1080/21681015.2016.1172124.
Bressanelli, G., Adrodegari, F., Perona, M., & Saccani, N. (2018). Exploring how usage-focused business models enable circular economy through digital technologies. *Sustainability, 10*(3). https://doi.org/10.3390/su10030639.
Brown, P., Bocken, N., & Balkenende, R. (2019). Why do companies pursue collaborative circular oriented innovation? *Sustainability, 11*, 1–23. https://doi.org/10.3390/su11030635.
Cainelli, G., De Marchi, V., & Grandinetti, R. (2015). Does the development of environmental innovation require different resources? Evidence from Spanish manufacturing firms. *Journal of Cleaner Production, 94*, 211–220. https://doi.org/10.1016/j.jclepro.2015.02.008.
Cervellon, M. C., & Wernerfelt, A. S. (2012). Knowledge sharing among green fashion communities online: Lessons for the sustainable supply chain. *Journal of Fashion Marketing and Management., 16*, 176. https://doi.org/10.1108/13612021211222860.

Cezarino, L. O., Liboni, L. B., Oliveira Stefanelli, N., Oliveira, B. G., & Stocco, L. C. (2019). Diving into emerging economies bottleneck: Industry 4.0 and implications for circular economy. *Management Decision.* https://doi.org/10.1108/MD-10-2018-1084.

Chen, D., Heyer, S., Ibbotson, S., Salonitis, K., Steingrímsson, J. G., & Thiede, S. (2015). Direct digital manufacturing: Definition, evolution, and sustainability implications. *Journal of Cleaner Production, 107*, 615–625. https://doi.org/10.1016/j.jclepro.2015.05.009.

Coreynen, W., Matthyssens, P., & Van Bockhaven, W. (2017). Boosting servitization through digitization: Pathways and dynamic resource configurations for manufacturers. *Industrial Marketing Management, 60*, 42–53. https://doi.org/10.1016/j.indmarman.2016.04.012.

Darnall, N., Jolley, G. J., & Handfield, R. (2008). Environmental management systems and green supply chain management: Complements for sustainability? *Business Strategy and the Environment, 17*(1), 30–45. https://doi.org/10.1002/bse.557.

De Jesus, A., Antunes, P., Santos, R., & Mendonça, S. (2019). Eco-innovation pathways to a circular economy: Envisioning priorities through a Delphi approach. *Journal of Cleaner Production, 228*, 1494–1513. https://doi.org/10.1016/j.jclepro.2019.04.049.

De Marchi, V., & Di Maria, E. (2019). Environmental upgrading and suppliers' agency in the leather global value chain. *Sustainability, 11*(23), 6530. https://doi.org/10.3390/su11236530.

De Marchi, V., Di Maria, E., Krishnan, A., & Ponte, S. (2019). Environmental upgrading in global value chains. In S. Ponte, G. Gereffi, & G. Raj-Reichert (Eds.), *Handbook on global value chains* (pp. 310–323). Cheltenham: Edward Elgar.

De Marchi, V., Di Maria, E., & Ponte, S. (2013). The greening of global value chains: Insights from the furniture industry. *Competition & Change, 17*(4), 299–318.

De Marchi, V., & Grandinetti, R. (2013). Knowledge strategies for environmental innovations: The case of Italian manufacturing firms. *Journal of Knowledge Management, 17*(4), 569–582. https://doi.org/10.1108/JKM-03-2013-0121.

De Sousa Jabbour, A. B. L., Jabbour, C. J. C., Foropon, C., & Filho, M. G. (2018). When titans meet – Can industry 4.0 revolutionise the environmentally-sustainable manufacturing wave? The role of critical success factors. *Technological Forecasting and Social Change, 132* (January), 18–25. https://doi.org/10.1016/j.techfore.2018.01.017.

Despeisse, M., Baumers, M., Brown, P., Charnley, F., Ford, S. J., Garmulewicz, A., et al. (2017). Unlocking value for a circular economy through 3D printing: A research agenda. *Technological Forecasting and Social Change, 115*, 75–84. https://doi.org/10.1016/j.techfore.2016.09.021.

Devinney, T. M., Auger, P., & Eckhardt, G. M. (2010). *The myth of the ethical consumer the myth of the ethical consumer.* New York: Cambridge University Press. https://doi.org/10.1177/0094306112438190m.

Ellen MacArthur Foundation. (2019). *Artificial intelligence and the circular economy – AI as a tool to accelerate the transition.* Retrieved from https://www.ellenmacarthurfoundation.org/publications

Esposito, M., Tse, T., & Soufani, K. (2018). Introducing a circular economy: New thinking with new managerial and policy implications. *California Management Review, 60*(3), 5–19. https://doi.org/10.1177/0008125618764691.

Geissdoerfer, M., Savaget, P., Bocken, N. M. P., & Hultink, E. J. (2017). The circular economy – A new sustainability paradigm? *Journal of Cleaner Production, 143*, 757–768. https://doi.org/10.1016/j.jclepro.2016.12.048.

Ghisellini, P., Cialani, C., & Ulgiati, S. (2016). A review on circular economy: The expected transition to a balanced interplay of environmental and economic systems. *Journal of Cleaner Production, 114*, 11–32. https://doi.org/10.1016/j.jclepro.2015.09.007.

Hazen, B. T., Mollenkopf, D. A., & Wang, Y. (2017). Remanufacturing for the circular economy: An examination of consumer switching behavior. *Business Strategy and the Environment, 26* (4), 451–464. https://doi.org/10.1002/bse.1929.

Hopkinson, P., Zils, M., Hawkins, P., & Roper, S. (2018). Managing a complex global circular economy business model: Opportunities and challenges. *California Management Review, 60*(3), 71–94. https://doi.org/10.1177/0008125618764692.

Horbach, J., & Rammer, C. (2019). Circular economy innovations, growth and employment at the firm level: Empirical evidence from Germany. *Journal of Industrial Ecology.* https://doi.org/10.1111/jiec.12977.

Huberty, M. (2015). Awaiting the second big data revolution: From digital noise to value creation. *Journal of Industry, Competition and Trade, 15*(1), 35–47. https://doi.org/10.1007/s10842-014-0190-4.

Jay, J., & Gerard, M. (2015). Accelerating the theory and practice of sustainability-oriented innovation. *MIT Sloan Research Paper.* https://doi.org/10.2139/ssrn.2629683.

Lacy, P., & Rutqvist, J. (2016). *Waste to wealth: The circular economy advantage.* New York: Palgrave Macmillan. https://doi.org/10.1057/9781137530707.

Moraga, G., Huysveld, S., Mathieux, F., Blengini, G. A., Alaerts, L., Van Acker, K., et al. (2019). Circular economy indicators: What do they measure? *Resources, Conservation and Recycling, 146*, 452–461. https://doi.org/10.1016/j.resconrec.2019.03.045.

Mylan, J., Geels, F. W., Gee, S., McMeekin, A., & Foster, C. (2015). Eco-innovation and retailers in milk, beef and bread chains: Enriching environmental supply chain management with insights from innovation studies. *Journal of Cleaner Production, 107*, 20–30. https://doi.org/10.1016/j.jclepro.2014.09.065.

Pittaway, L., Robertson, M., Munir, K., Denyer, D., & Neely, A. (2004). Networking and innovation: A systematic review of the evidence. *International Journal of Management Reviews, 5–6*(3–4), 137–168. https://doi.org/10.1111/j.1460-8545.2004.00101.x.

Porter, M. E., & Heppelmann, J. E. (2015). How smart, connected products are transforming companies. *Harvard Business Review, 93*, 91–114.

Rajput, S., & Singh, S. P. (2019). Connecting circular economy and industry 4.0. *International Journal of Information Management, 49*(March), 98–113. https://doi.org/10.1016/j.ijinfomgt.2019.03.002.

Roblek, V., Meško, M., & Krapež, A. (2016). A complex view of industry 4.0. *SAGE Open, 6*(2), 1–11. https://doi.org/10.1177/2158244016653987.

Saidani, M., Yannou, B., Leroy, Y., Cluzel, F., & Kendall, A. (2019). A taxonomy of circular economy indicators. *Journal of Cleaner Production., 207*, 542. https://doi.org/10.1016/j.jclepro.2018.10.014.

Seuring, S., & Müller, M. (2008). From a literature review to a conceptual framework for sustainable supply chain management. *Journal of Cleaner Production, 16*(15), 1699–1710.

Stock, T., & Seliger, G. (2016). Opportunities of sustainable manufacturing in industry 4.0. *Procedia CIRP, 40*, 536–541. https://doi.org/10.1016/j.procir.2016.01.129.

Tseng, M. L., Tan, R. R., Chiu, A. S. F., Chien, C. F., & Kuo, T. C. (2018). Circular economy meets industry 4.0: Can big data drive industrial symbiosis? *Resources, Conservation and Recycling, 131*, 146–147. https://doi.org/10.1016/j.resconrec.2017.12.028.

Ünal, E., & Shao, J. (2018). A taxonomy of circular economy implementation strategies for manufacturing firms: Analysis of 391 cradle-to-cradle products. *Journal of Cleaner Production, 212*, 754–765. https://doi.org/10.1016/j.jclepro.2018.11.291.

Unruh, G. (2015). The killer app for 3D printing? The Circular Economy. *MITSloan Management Review*, 17–19. Retrieved from http://sloanreview.mit.edu/article/the-killer-app-for-3d-printing-the-circular-economy/?utm_source=feedburner&utm_medium=feed&utm_campaign=Feed%3A+mitsmr+%28MIT+Sloan+Management+Review%29

Unruh, G. (2018). Circular economy, 3D printing, and the biosphere rules. *California Management Review, 60*(3), 95–111. https://doi.org/10.1177/0008125618759684.

Webster, K., & MacArthur, E. (2017). *The circular economy: A wealth of flows* (2nd ed.). Cowes: EllenMacArthur Foundation.

Yeo, N. C. Y., Pepin, H., & Yang, S. S. (2017). Revolutionizing technology adoption for the remanufacturing industry. *Procedia CIRP, 61*, 17–21. https://doi.org/10.1016/j.procir.2016.11.262.

Industry 4.0: New Paradigms of Value Creation for the Steel Sector

Laura Tolettini and Claudia Lehmann

Abstract Industry 4.0 has hugely affected the investment decisions of many industrial companies, including small and medium-sized enterprises. Thanks to national government initiatives, incentives and subsidies have supported investments in innovative and digital technologies, giving also a positive impulse in the national economic growth. The greatest advantages of Industry 4.0 are the empowerment of sustainable growth, thanks to the optimization of resources, and the integration of the supply chain. Nevertheless, Industry 4.0 challenges industries on the organizational level, concerning the integration of these technologies with existing ones and the management of digital tools by dedicated qualified people. Steel industries have also absorbed the challenges and the advantages of Industry 4.0. Digital technologies are affecting the business model and supply chain of steel producers, pushing them to integrate themselves with their business partners. Industry 4.0 is definitely an opportunity of sustainable growth for the steel sector.

1 Introduction: Development and Initiatives of Industry 4.0 in Europe and Worldwide

Industry 4.0 (I40) is a national initiative jointly launched in 2011 by the German Ministry of Education and Research (Bundesministerium für Bildung und Forschung—BMBF) and the Ministry for Economic Affairs and Energy (Bundesministerium für Wirtschaft und Energie—BMWi). Chancellor Angela Merkel recognized the urging necessity to create a coherent governmental strategy

---

L. Tolettini (✉)
Feralpi Holding S.p.A., Lonato del Garda, Italy
e-mail: laura.tolettini@it.feralpigroup.com

C. Lehmann
CLIC Center for Leading Innovation and Cooperation, HHL Leipzig Graduate School of Management, Leipzig, Germany
e-mail: claudia.lehmann@hhl.de

© Springer Nature Switzerland AG 2020
M. Bettiol et al. (eds.), *Knowledge Management and Industry 4.0*, Knowledge Management and Organizational Learning 9,
https://doi.org/10.1007/978-3-030-43589-9_8

in order to integrate the online world with the world of industrial production and giving the German companies the chance to be competitive on the new globally digitalized market. In this way, the German Government immediately allocated a fund of €200 million, in order to serve the High Tech 2020 Strategy (Klitou et al., 2017a, 2017b, 2017c, 2017d, 2017e, 2017f). At the same time, Germany created a national I40 platform in order to put all the players involved (companies, research centers, and consultants) in constant communication, also with the goal to support the development of the needed skills, especially in the STEM (Science, Technology, Engineering, and Mathematics) field (Bundesregierung, 2015).

Industry 4.0 is considered the fourth industrial revolution. The first industrial revolution was ignited by mechanization, through the employ of water and steam power. In the second revolution, electricity enabled assembly lines and mass production, and it became the third revolution with the further development of automation and computers. Industry 4.0 is a step further, using cyber-physical systems in order to integrate the physical and the digital world, leading to smart autonomous industrial systems (Marr, 2018).

Via embedded sensors and actuators operating in a wireless network system, cyber-physical systems (CPS) interconnect objects, by exchanging a huge amount of data in a very short time. The physical object acquires a so-called digital twin, which is the representation of this object in the virtual world (Boschi, De Carolis, & Taisch, 2017). CPS are more diffused in daily life than one could imagine: 90% of all microprocessors work in embedded systems, which have become a critical point for the competitive advantage of companies, being able to network human and objects with strategic information (BMBF, 2018). CPS are the key for integration and interconnection promised by Industry 4.0 enabling technologies. CPS lead machine to make autonomous decision and to influence and change the environment where they are working. They enable the configuration not only of smart factories but also of smart environment and supply chains to which a company is interconnected (Pflaum et al., 2014).

There are nine key enabling technologies (KET) of Industry 4.0, as the Italian Ministry for the Economic Development (MISE) has recently pointed out (MISE, 2018):

1. Advanced Manufacturing Solutions: collaborative, interconnected, and immediately reprogrammable robots, which can work in the same environment of human beings.
2. Additive Manufacturing: 3D (three-dimensional) printers which work through dedicated software in order to produce customized objects.
3. Augmented Reality (AR): the integration of virtual reality in the world of production in order to facilitate daily tasks or training at the machine.
4. Simulation: software enabling the virtual representation of different scenarios among connected machines in order to optimize manufacturing processes.
5. Horizontal and vertical integration: thanks to the rapid and online data exchange, the supply chain can be fully integrated, from producer to consumer.
6. Industrial Internet or Internet of Things (IoT): thanks to CPS, human, machines, and products are constantly interconnected and can adapt fast to unexpected environmental changes.

7. Cloud Computing: hardware is virtualized and information is stored in open platforms, which enables almost an infinite space for memorization, at lower costs and at higher speed.
8. Cybersecurity: it creates a secure environment for machines and software in order to exchange sensitive information and avoid dangerous stealing of data from internal and external counterparts.
9. Big Data Analytics: millions of data can be exchanged in milliseconds. Using and interpreting data have become an essential source of competitive advantage, independently from the sector where companies are working.

Through these technologies, the fourth industrial revolution leads companies to integrate all steps of production and all levels of interaction among the different departments, but at the same time to be more networked with their customers and suppliers. In this sense, two other technologies could be added to the traditional list of KET mentioned before: artificial intelligence (AI) and Blockchain.

Concerning artificial intelligence, the symbiosis among humans and machines becomes so deep that machine learns to interact like humans from humans themselves. Through complex algorithms, machines gain new knowledge of the external world (Rouse, 2018), and they are able to automatically adjust production to anticipated availability (Maniyka, 2011). Artificial intelligence will further boost the ability of Industry 4.0 technologies to increase productivity and product quality (Berg, 2019).

The Blockchain is a digitized, decentralized public ledger of all cryptocurrency transactions. A block is created to represent a requested transaction and is sent to all nodes of the network in order to be validated. Once it is validated, the block is added to the existing Blockchain and cannot be removed. Validators can keep track of all transactions chronologically, without the support of a centralized authority. Blockchain could be very useful in the validation of transactions concerning sensitive information also in the industrial sector, e.g., related to the creation of contracts and currency transactions (Peterson, 2018). It is very interesting how Industry 4.0 is also affecting the way knowledge is transmitted and transformed digitally, in order to enable a faster and more secure communication. The revolutionary aspect of digitization is that knowledge is generated from a source and controlled by different participants of a network in a diffusive way, without the necessity of a centralized authority. For example, this has already happened with Wikipedia, creating a system of verification and authorization of participated knowledge by the support of an entire community of singular users. Digital technologies empower a networked community to exchange and validate notions, in an integrated system.

The integration force of the technologies of Industry 4.0 leads to appealing advantages and opportunities for all industrial sectors: optimization of production and maintenance costs and of consumption of resources; higher product quality through a better control of production; flexibility and customization of manufacturing; and new market and supply chain opportunities (European Parliament, 2016). Industry 4.0 puts the customer at the center of the manufacturing process and delegates operative productions tasks to machines and robots (Pareekh & Tanmoy, 2017).

Some important challenges are also driven by Industry 4.0: costs for the integration of new and existing technologies; requalification of workforce and the search for new interdependent soft and technical skills; and protection of intellectual property and data privacy and security (European Parliament, 2016). Certainly, Industry 4.0 implies a huge organizational effort from managers. The successful integration of innovative technologies with the existing industrial reality forces managers to align the organization to a vision and to clearly define a realistic roadmap to fulfill with the support of both internal (employees) and external (customers and suppliers) stakeholders (Brunelli, Lukic, Milon, & Tantardini, 2017). Industry 4.0 challenges companies on the topic of innovation and knowledge management. Since digitization enables the collection of data from many heterogeneous sources, managers have to train themselves and their staff to interpret these data by thinking on a multidisciplinary level and by integrating operative and strategic information to create a competitive advantage for their company. The investment in innovative technologies and the modernization of plants and infrastructure increases the value of a company, its knowledge, and expertise. Nevertheless, the process of technological integration requires a structured way to manage innovation in the daily work, so that innovation remains not only a sporadic phenomenon of a genial idea, but an incremental process to support sustainable competitive advantage.

Indeed, Industry 4.0 represents a fundamental chance for the European Union (EU) in order to gain sustainable competitive advantage in the globalized market. Many national initiatives started in different regions of the EU. These initiatives can be linked by a common strategy: they supported not only the creation of innovative startup from scratch, but they wanted to ignite new potentials in the traditional sectors and to create advanced technological poles with the help of universities and research centers. This enriches the innovation knowledge of an industrial community, by integrating the perspectives of different partners (high-tech and traditional companies, universities, governmental institutions). Each European region has tried to specialize in a particular innovative sector, but at the same time, they fostered the exchange of best practices both on a European and on an international level.

One year after the German initiative, the Italian Government installed a budget of €40 million, in order to create synergies among industries, research centers, and associations (Klitou et al., 2017a, 2017b, 2017c, 2017d, 2017e, 2017f). Later, the Italian Ministry for Economic Development (MISE) incremented the fund up to €13 billion for the period of 2017–2020 to sustain enterprises and startups in R&D and investments in Industry 4.0, also through financial measures like super-amortization and tax relief (MISE, 2018).

In France, Industry 4.0 took the name of *Industrie du Futur*, with the initiative launched in 2015, with a budget of €10 billion (Klitou et al., 2017a, 2017b, 2017c, 2017d, 2017e, 2017f). The goal was to integrate the entire production process with smart digital technologies, in order to offer faster and tailored designed products and services to customers. The initiative involved nine industrial areas: (1) transport, (2) smart objectives, (3) new resources, (4) medicine, (5) digital security, (6) sustainable cities, (7) data economy, (8) smart food production, and (9) eco-mobility (Klitou et al., 2017a, 2017b, 2017c, 2017d, 2017e, 2017f).

At the end of 2010, the Government in the United Kingdom (UK) provided additional funding for more than £200 million for the period 2011–2015, in order to foster high-value manufacturing centers, the so-called Catapult Centers, focused on some strategic innovation topics, like digital technologies, renewable energy, future cities, and future transportation system (Hauser, 2014).

In 2016, the Spanish Government granted €97.5 million for innovative and research projects in industries, €68 million for ICT (Information and Communication Technologies) companies, and €10 million for innovative clusters under the label of *Industria Conectada* (Klitou et al., 2017a, 2017b, 2017c, 2017d, 2017e, 2017f).

Many other Industry 4.0 initiatives fostered in those years across Europe, to quote some more: Smart Industry (Holland, 2014, €25 million), Production 2030 (Sweden, 2013, €50 million), MADE (Manufacturing Academy of Denmark, 2014, €50 million), and Produtech (Portugal, 2017, €4.5 billion) (Klitou et al., 2017a, 2017b, 2017c, 2017d, 2017e, 2017f).

In the world, Industry 4.0 covered different aspects. In the United States of America (USA), President Obama dedicated USD $1.3 billion to production research in 2013, USD $1 billion for the National Network for Manufacturing Innovation, and USD $300 million for a Wireless Innovation Fund (Sargent, 2013).

In the USA, Industry 4.0 took the name of "Smart Production" or "Industrial Internet," and it is more focused on the industrial and consumer employ of Big Data Analysis in order to create new business models ("smart services") (Kagermann, Anderl, Gausemeier, Schuh, & Wahlster, 2016). In 2014, some famous hi-tech companies, like Microsoft, IBM, and Intel, founded the Industrial Internet Consortium, in order to put together industrial leaders, research centers, universities, and the government to build best practices and create framework for the implementation of IoT technologies. This coalition is collaborating with the German platform *Industrie 4.0* to develop compatible architectures for the industrial interoperability in order to foster standardized knowledge in the high-tech field (Industrial Internet Consortium, 2019).

With its plan "Made in China 2025," China will create 15 innovation centers by 2020, and 40 by 2040. It wants to increase the national production of Industry 4.0 core components and products to 40% by 2020 and 75% by 2025. Priority sectors have been identified for this technological development: new advanced information technology; automated machine tools and robotics; aerospace and aeronautical equipment; maritime equipment and high-tech shipping; modern rail transport equipment; new-energy vehicles and equipment; power equipment; agricultural equipment; new materials; and biopharma and advanced medical products (Lee, 2015). Through digitization and innovation, China wants to boost its economic paradigm, by achieving manufacturing excellence, by increasing investments in R&D, and by concretely revising demand planning, sustaining more efficient and flexible organizations (Eloot, Huang, & Lehnich, 2013). Between 2016 and 2018, the majority of Chinese investments in Europe were acquisitions in strategic technological fields and 60% of them were made by the government itself (Santoro, 2019). We can observe how gradually the innovation paradigm of China started to support a more sustainable economy, also concerning environmental issues.

High-value manufacturing in India is supported by the initiative Make in India, launched in 2014 and led by the Department of Industrial Policy and Promotion (DIPP). The initiative wants to eliminate bureaucracy and lack of transparency, to attract even more foreign investments in 25 different industrial sectors (Make in India, 2019).

In 2014, the Russian President, Vladimir Putin, announced the implementation of the National Technology Initiative (NTI), to make Russia a technological leader in the following sectors in the next 15 years: energy (distributed power from personal power to smart grid and smart city), security (new personal security systems), finance (decentralized financial systems and currencies), and neurological research (with use of AI) (ASI, 2019).

It is important to mention two other Asian industrialized leading countries concerning initiatives of Industry 4.0, where there are some important steel producers. In the following paragraphs of this chapter, we will see how these industries are not only investing in technological innovation for their traditional operating activities, but they are trying to anticipate and redesign the future of their sector. For these industries creating spaces for creativity and innovation to gain new knowledge and competitive advantage toward their competitors is fundamental. The Manufacturing Industry Innovation 3.0 plan is part of South Korea's Creative Economy Initiative. The goal of the government is to have 30,000 smart factories by 2025 (Son, 2018). The Industrial Value Chain Initiative (IVI) is the Japanese initiative to promote collaboration among industries on topics related to digital technologies, also expanded to international collaborations. Working groups are especially engaged to facilitate the industrial use of CPS through a standard interface and of IoT for smart maintenance (IVI, 2019).

2 Development of Industry 4.0 in the Steel Sector

The steel industry lies at the heart of the industrial economy worldwide. Its origin is recognizable in the mid-1850s, with the first mass production at inexpensive costs (Saville, 2019).[1] It employs more than six million people around the world directly and 40 million people indirectly in the supply chain. In 2018, it produced a total amount of 1.8 billion tons crude steel, and in 2017, it created US$500 billion value added (World Steel Association, 2019). Steel industry is the backbone of the European economy, too, with 330,000 highly skilled people directly employed and 170 million tons of produced crude steel per year at more than 500 steel production sites across 23 EU member states (Eurofer, 2019).

---

[1]The first inventor of mass cheap steel production was the British Henry Bessemer (1813–1898). In 1856, he was able to discover a new method of producing steel, in large slag-free workable ingots (Saville, 2019).

Steel is employed in a multitude of different scopes in everyday life (e.g., buildings and infrastructure, automotive, mechanical, and electrical equipment, domestic appliances) and is a recyclable material. It can be reused 100% as raw material in the production method of the Electric Arc Furnace (EAF); on average, new steel products contain 37% recycled steel (World Steel Association, 2019).

Steel industries are facing determinant challenges for their existence. Being cyclically affected by overcapacity, they must always search for innovative ways of sustainable competitive advantage. In its last outlook, the Organisation for Economic Co-operation and Development (OECD) expressed its concern about a new cycle of overcapacity in the steel sector, driven by the lower growth of the global economy (OECD, 2019).

Steel industry has to cope with these fundamental challenges concerning its sustainable growth:

- Environmental sustainability: reducing Co_2 emissions and optimizing energy consumption, by also using renewable sources. In the last few decades, steel industry reduced Co_2 emissions by 60%, and they pursue the further ambitious goal of zero emissions until 2050 (Kerkhoff, 2019). One another important point is enhancing the use of co-products, generated by the output of the steel process (like slags and dusts). Nearly 100% of steel industry output material can be reused (World Steel Association, 2018).
- Social sustainability: the acceptance by the local community is fundamental for the successful development of the steel sector. At the same time, social sustainability is intended to guarantee the safety of the employees at work.
  In 2004, the World Steel Association started the initiative of sustainable steel: eight annual indicators, also concerning social performance, to which steel industries can report voluntarily (World Steel Association, 2018).
- Innovation sustainability: steel must remain competitive as material toward alternative products. In 2017, the steel industry invested 5.9% of revenue projects concerning the development of new steel types (World Steel Organization, 2019). Innovation in the supply chain is also indispensable in order to keep the pace of the growth of the other industrial sectors.

Industry 4.0 represents a great chance for steel industries to keep growing sustainably. Since 2013, the applications of Industry 4.0 have been put as a priority in the RFCS—Research Fund for Coal and Steel—the research program for steel industries.

The huge potential of I40 consists in increasing the optimization of resources and reducing the impact on the environment and on the local community. This is possible through the integration of all production steps and to the opportunity to predict and avoid process failures with the use of machine learning and big data analysis. Final products can be tracked intelligently up to the customer and can be matched precisely to process failures in case they are defective. Sensors and radars can measure and control furnace parameters and warn in case of dangerous operations.

Machines can be networked together in platforms, where production operators can use online data and react supportively. Drone technology can be useful to inspect areas difficult to reach and to classify and value raw material, like scrap and alloys (Janjua, 2018). Already at the end of the 1990s, first applications of AI and neural networks were already used in statistical analysis and in results previsions for technical processes.

One crucial point of I40 for steel industries is the opportunity to be more integrated in the supply chain and to offer more customized products and services, also in the case of the commodity market. If customers want to enable data exchange, producers can understand the employ of their final products better and assist customers in their needs more effectively (Kepmf, 2018). This leads to new business models, where steel producers can offer not only physical products but also additional services.

The digital transformation challenges steel industries in their established organization: new technologies have to be integrated in existing plants and new competences are required in the technical staff, which has to cope with digital, technical, and organizational problems (Janjua, 2018).

In Germany, one of the biggest steel producer in Europe, two interesting studies were made about digital transformation in the steel sector. They can be useful in recognizing the practical patterns behind a possible hype and observing how Industry 4.0 is affecting innovation and knowledge management in the steel sector.

In 2017, the German Association for Steel Industries (Wirtschaftsvereinigung Stahl) commissioned IW Consulting to analyze the real application of digital transformation in the sector, based on the results of the IW-Zukunftspanel with about 220 customers and 60 companies of the steel sector.

Results underline these nine following trends related to Industry 4.0 in the steel business (IW Consult, 2017):

1. Increased innovation competences: producers can be earlier bound into the development phase of the products of their customers. This enlarges the technological and commercial knowledge of steel industries, offering new market opportunities.
2. Larger product portfolio: steel products can be integrated with digital components in order to track the product itself and to tell features about the product.
3. Higher flexibility of the supply chain: thanks to data exchange, suppliers are more networked to their customers and can assist them immediately in after-sale activities.
4. Hybridization: product portfolio can be expanded thanks to ideas gained by stakeholders. New services and business models can emerge, in order to be more competitive on the market.
5. Virtualization: production can be virtually simulated in order to avoid waste of time and resources.
6. Shorter lead times: time to market becomes shorter thanks to the speed of the integrated processes.

The supply chain of steel industries includes a variety of stakeholders. Steel producers do not always have direct contact with their suppliers and customers. Suppliers of steel producers can be divided into four categories: suppliers of raw materials and additives (e.g., scrap, iron ore, alloys, lime, coal); suppliers of semi-finished products (tubes, wires, bars, ingots, and billets); and suppliers of services (transportation, energy, maintenance) and of equipment (spare parts, new parts of the plant) (IW Consult, 2017). In the same way, customers of steel producers can be categorized into those four groups: construction industry (35%); automotive (18%); mechanical engineering (14%); metal goods (14%); tubes (13%); domestic appliances (3%); other transport (2%); and miscellaneous (1%) (World Steel Association, 2019).[2]

In February 2019, the Fraunhofer Institute for Production Technique and Automation (IPA) did a study on the grade of reception of Industry 4.0 in the biggest customers of steel producers: metal and steel trading companies and service centers.

An online questionnaire was sent to 400 German companies representing the steel trading market, with a response rate of 16.5%. Afterward deeper qualitative expert interviews were fulfilled (Schumpp, Birenbaum, & Schneider, 2019). Answers display a trend of introducing two levels of digital integration within steel producers: a first level concerning the simplification of administrative documents (invoices, orders, contracts) with the help of tools like EDI (electronic data exchange) and a second level of integration, using web platforms and mobile applications, taking the example from the consumer market. Digital transformation is often driven by the customer or supplier, who challenges the rest of the supply chain to adapt to new digital technologies (Schumpp et al., 2019). Industry 4.0 supports the integration of competences and knowledge among all stakeholders of the supply chain, affecting the way of developing innovation. Through Industry 4.0, innovation is a result of collaboration, leading to the approach of open innovation.

3 Analysis of a Company Case: Feralpi Group

3.1 Presentation of Company Case and Historical Development of the Company

Feralpi Group is an international family company with the core business of steel for construction industry. With an output of 2.5 million of crude steel, 3.3 million of final products, and a turnover of €1.32 billion, Feralpi is one of the most successful Italian and European steel producers, counting more than 1500 employees. Nowadays, more than 60 % of turnover is generated abroad. Feralpi is a significant company case to analyze. Its history shows the development of a solid national family company leading to the creation of a multinational group, principally based in Italy and in Germany. Feralpi case offers the opportunity to analyze the technological evolution and acquisition of Industry 4.0 knowledge and competences in the two

---

[2]Percentages of consumption intended as percentage of steel consumption for each customer category on the total apparent steel consumption.

most important nations for steel production in Europe, Italy, and Germany. Feralpi is an interesting and successful example of how the vision of its entrepreneurs made possible to integrate best practices from two different contexts, Italy and Germany, even though sometimes their mentality and culture can be quite different.

Feralpi Group was founded in 1968 in Lonato del Garda by Carlo Nicola Pasini and some other partners, including the families Tolettini and Leali. The site of Feralpi Siderurgica SpA in Lonato offered a strategic industrial position next to the railway and the motorway. Until 1973, Feralpi made different technological investments, concerning two highly automated rolling mills and a continuous billet-casting machine. The technical development continued, with the construction of an additional steel plant in Calvisano, 20 km from Lonato, in order to serve the central plant. Technical innovation was always an important strategic point for Feralpi to be distinguished from competitors. In 1988, Feralpi was the first Italian commodity steel producer to obtain a patent for the Tempcore process for the production of rebars.

In the 1990s, when the European Community began to restructure the steel sector distributing subsidies to close and aggregate factories, Feralpi Group started its internationalization process: production units in Hungary (Ozd and Budapest) and in the Czech Republic (Kralupy) were acquired. In 1992, a very strategic merging was made: from a former state-owned company of 10,000 employees, the ESF Elbe-Stahlwerke Feralpi was founded. The site is in Riesa, Eastern Germany, offering a very competitive market position, both for the main raw material (scrap) and for final products. In 1999, Feralpi entered the Romanian market, acquiring a 50% equity interest in Ductil Steel Sa, manufacturing for construction industries, too. This adventure, finished in 2007, when Feralpi sold its stake, due to significant cultural and market challenges. Today, Feralpi has a participation in Beta, a Romanian company, which is active in the carpentry industry. In 2009, the verticalization process grew in importance, with the acquisition of Nuova Defim S.p.A. (Como), producer of electro-welded meshes for industrial use and wire fencing and gates for special applications. In 2012, Feralpi acquired Orsogril, a subsidiary of Nuova Defim, which produces steel gratings for application in building and architectural projects. In 2013, the group set up Feralpi Algérie, a commercial company based in Oran, to serve the increasingly growing North African market. In 2014, Feralpi started to focus on differentiation, investing in the noble steel production, with Caleotto S.p.A. (Lecco), a rolling mill for the production of top-grade wire rod, in a joint venture with a partner of the noble steel sector. In 2015, Feralpi made a new joint venture, entering Cogeme Steel, in order to come closer to the final customer, with special products for private construction industry, automotive, and engineering. In the same year, it took a stake in Presider and Metallurgica Piemontese Lavorazioni (Torino), specialized in pre-processing and on-site installation of rebar and steel beams.

On May 2016, the group made its latest acquisition, with Profilati Nave Spa (Brescia), an Italian producer in special profiles, in order to enlarge its potential in the noble steel industry (Feralpi Holding S.p.A., 2019) (In Fig. 1 we show the company organization).

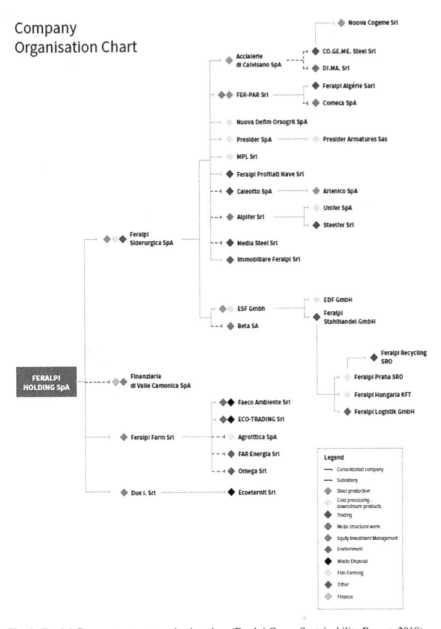

Fig. 1 Feralpi Group company organization chart (Feralpi Group Sustainability Report, 2018)

3.2 Presentation of Research Scope and Method

As we see from the development of the group, Feralpi strategy is based on three pillars: internationalization, verticalization, and differentiation, which enabled

Feralpi to gain more solid knowledge and experience of the market and of innovative technologies. The strategy of internationalization meant to come from commercial relationships with the main foreign partner (Germany) to create a production unit abroad to serve directly the market and to exploit synergies with next markets (Eastern Europe: Czech Republic and Hungary). Verticalization had the goal to make production costs more efficient and try to come next to the final costumer (construction industry), by transforming the semifinished product (wire rod) in welded meshes and stretched coils. Moreover, the participation in control or in a joint venture in two significant steps of the supply chain (scrap collection with Media Steel and steel reshaping with Presider) gave Feralpi the opportunity to know the market better and acquire new competences in collaboration with its suppliers and customers. Differentiation gave the opportunity to explore a new segment of the market like the one of higher quality of steel with the participation in Caleotto S.p. A., exploiting the consolidated know-how in steel making and at the same time enlarging their technical and commercial experience to more tailor-made products for customers (wire-drawing mills).

Nevertheless, a solid pillar behind this strategy has always been the technological development. The technological development of Feralpi reveals a continuous innovation effort in order to offer the best product and the best service in a commodity-dominated market. Continuous investments also in time of crises led Feralpi to be prepared for the changes of the market and to understand the needs of the customers better.

Today, Industry 4.0 is driving the technological decisions of Feralpi Group. In order to outline the strategy, we interviewed managers of the headquarters active in the field of Industry 4.0 and sustainability. Qualitative problem-centered interviews were conducted together with Università degli Studi di Padova, Laboratory of Digital Manufacturing, led by Prof. Eleonora Di Maria, and supported by Valentina De Marchi.

The Laboratory for Digital Manufacturing (LMD) is a university initiative with the goal to dig deeper in the concrete transformation of digital technologies and Industry 4.0 on the Italian manufacturing companies. Its research group analyzes the consequences of digital technologies on strategies and business models of Italian producers (LMD, 2019). One of its research area is to study the impact of Industry 4.0 on the circular economy and on the sustainability patterns of Italian industries. In 2018, LMD conducted an important enquiry together with Legambiente[3] on this topic, underlining how bureaucratic constraints hinder companies to fulfill important best practices in terms of sustainability goals (Di Maria, De Marchi, & Blasi, 2018).

An interesting collaboration started with LMD and CLIC, Center for Leading Innovation and Cooperation of the HHL Leipzig Graduate School of Management.

---

[3]Legambiente is an Italian nonprofit association which has the goal to protect the environment and to raise awareness on the human impact on climate change and environmental transformations. Founded in 1980, it collaborates concretely with industries and universities in order to make scientific studies, which can help decision and strategies in different business fields (Legambiente, 2019).

CLIC is focused on applied innovation in companies and it researches on the technological application of Industry 4.0 in companies, by collaborating with international universities and participating in different European and international projects (CLIC, 2019).[4] We decided together to analyze the case of Feralpi Group and we received very open collaboration.

In May 2019, we made our interviews in Feralpi. We interviewed the Chairman, the Managing Director, the person responsible for Corporate Social Responsibility (CSR), the person responsible for the environment, and the person responsible for the application of Industry 4.0 technologies of Feralpi Group. Interviews were conducted vis-à-vis and lasted one hour on average.

We interviewed managers on these fundamental questions in order to make a snapshot of the company on its strategy, Industry 4.0, and sustainability:

1. How do you define your competitive advantage and your business model?
2. Which technologies of Industry 4.0 are adopted in your company and why?
3. Which impact had those technologies on your organization, on your business partners, and on your business model?
4. Which meaning has sustainability for your company and which challenges did you face in order to fulfill your sustainable goals?
5. How did Industry 4.0 support your sustainability goals?

We transcribed all interviews and analyzed their results, by aggregating the responses concerning the topics considered.

3.3 Results of the Research

The steel market is dynamic and influenced by national and international political decisions concerning production and commercial capabilities. In 2008, the steel sector was hit by an important crisis, after some positive years of expansion. Observing the two main markets where Feralpi is present, Germany and Italy, they had a different recovery, influenced also by the economical and industrial development of the two nations. Germany recovered earlier than Italy from the economic crisis of 2009, hence supporting the growth of the industrial sector, including steel production. Through its presence in Germany, Feralpi Group could counterbalance the weaker Italian market. Since 2009, Feralpi has grown constantly in Germany, while the Italian market reduced itself to a third of its capacity. The strategical decision to cope with the situation of the Italian national market was to start focusing

---

[4]The collaboration to analyze the case of Feralpi started with participation in the workshop "Creating Value through Manufacturing: Exploiting Industry 4.0 in a Circular Economy Framework," in March 2019. The workshop saw the participation of different European universities. In this circumstance, we presented the results of the research we conducted in August 2016, where we interviewed 56 protagonists of the steel sector in Germany and in Italy about the challenges and chances of Industry 4.0.

on diversification, by entering the niche market of specialties, which is a strategy still in development. Steel production for special steel and for commodity is very different from a technological and commercial point of view. In specialties, there are also market entry barriers, since customers are quite loyal to established market leaders. Feralp

i Stahl[5] serves mainly the national market, while the national Italian market is characterized by a high density of competitors in the same province and there is more necessity to export products in Europe or in other foreign countries, like Northern Africa and also Canada and the USA. In Europe, Switzerland, Spain, and Germany are very strong competitors, while internationally the market is constrained by the American duties and by the force of Turkey.

The internal commercial structure manages a network of agents and service centers, which have to transform and deliver products for the construction sites. A further relevant difference between the Italian and German market are payment conditions. In Italy, payment conditions are much longer than in Germany. In Italy there is more necessity to manage the credit situation of the customers, some of them being small companies, which do not always receive insurance cover.

The structure of suppliers of Feralpi is quite stable, characterized by 35% of regional suppliers. In Germany, scrap, the principal raw material for Feralpi, is supplied by big companies, while the market in Italy is more fragmented (Interview Managing Director Feralpi Holding, 2019c).

As we already saw in the paragraph concerning the presentation of the steel sector, the supply chain of steel industries like Feralpi is made of diverse players, with different sizes and origins. Core technical processes of Feralpi are the preparation, melting, solidification, and rolling of steel. Through the years, Feralpi has always tried to improve its core processes and to control the quality of its strategic raw material, in order to offer the best possible quality to its customers (Fig. 2 simplifies the supply chain of Feralpi).

One driving force for Feralpi, together with technological development, has always been to invest in sustainable growth. Even if it was not mandatory, Feralpi was the first company in its province to voluntarily engage in a sustainability balance, in 2006. The necessity was to communicate to the community all actions done in matter of sustainable goals (e.g., sustaining the local community and own employees, and protecting the environment), in order to gain credibility and authoritativeness from the own stakeholders (Fig. 3 displays the most strategic stakeholders of Feralpi).

Through the years, Feralpi improved its approach to map its processes according to the demanding standards of the Global Reporting Initiative (GRI).[6] The process is still going on, and in the latest steps, it should require to integrate sustainability objectives with the economical and production goals in the Managing by Objective

---

[5]Feralpi Stahl is the trading name for the four companies of Feralpi in Riesa: ESF Elbe-Stahlwerke Feralpi GmbH, EDF Elbe-Drahtwerke Feralpi GmbH, Feralpi Stahlhandel GmbH, and Feralpi Logistik GmbH.

[6]The GRI (Global Reporting Initiatives) is an independent international organization that helps businesses and governments understand and communicate their impact on sustainability issues (climate change, human rights, governance, and social well-being) (GRI, 2019).

Industry 4.0: New Paradigms of Value Creation for the Steel Sector 193

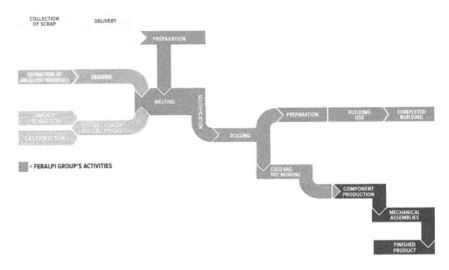

Fig. 2 Simplification of Feralpi supply chain (Feralpi Group Sustainability Report, 2018)

Fig. 3 Feralpi strategic stakeholders (Feralpi Group Sustainability Report, 2018)

(MBO) system of managers, as it already happens in some companies, champions of sustainability.

The sustainability engagement of Feralpi has its core force first in the engagement toward its employees. Safety is a central issue and a critical point for all steel companies. Safety is a global concept. It is intended as safety of its employees at work, including realization of measures of welfare and of prevention of professional diseases. If employees are healthy and feeling good, the company is becoming more successful.

Some years ago, Feralpi decided to do an internal survey both in Italy and in Germany, which is done every two years. The goal is to collect all information concerning the satisfaction of its employees. The survey is anonymous and is conducted and verified by a partner university. In this way, employees have the concrete opportunity to understand the sustainability engagement of Feralpi and, at the same time, Feralpi receives a concrete objective feedback on the measures done.

Digital technologies can support safety at work. There are currently systems, which can signal location of people inside the workplace, sending and receiving alarms in case of man down and displaying the shortest path to reach missing people.

Nevertheless, in the digital era, security is also intended as integrity and protection of company and personal data. In the time of cybersecurity, the reputational risk of companies is very high. In this sense the collaboration with trusted suppliers is a pillar in order to create customized plants and manage data in an efficient and sustainable way (Interview CSR Manager Feralpi Holding, 2019b).

The environmental impact is an extremely important issue concerning sustainability and the use of advanced technologies. Today the production paradigm is changing with the circular economy. Output material from the production can be transformed into by-products, giving the company the opportunity to explore new business and new markets. This is the case of Feralpi, concerning the reuse of slags. The black slag, coming from the melting process, is transformed from waste into a by-product ("Green Stone +2"), and certified according to UNI EN standards, which can be used in road construction and aggregates for concrete and bituminous mixtures (Interview Responsible for Environment Feralpi Siderurgica SpA, 2019d). Another case concerns the plants of Feralpi in Germany. Steam is captured from the electric arc furnace (ESF) and used by the town municipality to warm up buildings in winter. This model was replicated in Lonato last year and offers a clear example of integrated smart factory and smart city.

Linked to the topic of sustainable growth is the employ of Industry 4.0 technologies. Industry 4.0 fits in the continuous innovation strategy of the group, giving also the opportunity to exploit new technological possibilities.

For Feralpi Industry 4.0 must be concretely integrated in the everyday life of the production process. This implies not only a technological issue (functionally integrating new and existing technologies), but at the same time an organizational challenge, by spreading the digital culture in all levels and function of the managing and operative staff (Interview Managing Director Feralpi Holding, 2019c).

Considering the nine KET of Industry 4.0, Feralpi is generally investing in all of them with different scopes:

1. Advanced Manufacturing Solutions: labeling and sampling systems in the steel plant and rolling mill, which allow product traceability, and robots substituting human labor in dangerous activities.
2. Additive Manufacturing: at the moment, 3D printers are not so interesting for the business of Feralpi. They could be useful in case of production of prototypes or as additional service offered by business partners in order to supply spare parts faster.
3. Augmented Reality (AR): there are some analyses concerning the use of AR as support for e-learning training and for maintenance activities.
4. Simulation: since 2009, simulators are used from the melting to the rolling process, in order to calculate the perfect mix of raw materials and technical

parameters in order to improve the final product; monitoring systems are based on key performance indicators (KPI).
5. Horizontal and vertical integration: development of an e-business platform for final products with intelligent system of production planning.
6. Industrial Internet: advanced sensors and machine learning are used for predictive quality in order to come to "zero defects" and to be able to produce new steel qualities; in 2012, Feralpi had already introduced the basis of machine learning.
7. Cloud Computing: implementation of an ERP (Enterprise Resource Planning) system in cloud in order to connect different plants in remoting.
8. Cybersecurity: implementation of cybersecurity protocols and systems in order to protect software and machines.
9. Big Data Analytics: modeling systems based on KPI and on historical data in order to self-adapt the process to the best possible outcome.

In Feralpi, Industry 4.0 is driven by three main technological goals:

(a) Collection, evaluation, and interpretation of data: the smart factory leads to have numerous data available. The central issue is to find the most important data which can help build strategic economical and qualitative KPI. The technical and organizational issue is to be able to interpret the existing data and to be able to measure new data with specific technology. Another important factor is time. Being faster is a central point to gain competitive advantage. Data are synchronically collected and integrated from different sources, while repetitive parameters are synthetized in concise reports. The integration of various data to have also an economical interpretation is also an organizational and technical challenge.
(b) Advanced calculation systems: simulators are principally offered by machinery suppliers and they are tailor-made for the final customer.[7] Those systems start from a physical parameter; they are then correlated to other data and finally transformed in utility parameters.[8] There are different types of simulations. Process simulations for EAF (electric Arc furnace), LF (ladle furnace), and CM (casting machine) are mainly done by special consultants. Online simulations are both internally programmed and by external suppliers.
(c) Robotization: automation and robotization have always had a positive impact on manual strenuous work. They enhance safety at work and, at the same time, they guarantee a better integration of the process and of the traceability of the product.

(Responsible of Industry 4.0 applications Feralpi Holding interview, 2019e).

The approach to Industry 4.0 has two key elements for Feralpi: the centrality of human capital and sustainability. Machines are automated and self-adapting, but central decisions are made by human operators. Workers are a central pillar for Industry 4.0 in Feralpi: they are not simple executors of technology, but they are

---

[7]FEM: Finite element method is the most widespread method of simulation.

[8]For example, the physical parameter of temperature is transformed in an energy parameter in order to decide concrete actions to improve the performance of the furnace.

mediators between machines and the rest of the organization. Workers are active participants in the implementation and integration of Industry 4.0 technologies in the existing structure. They can support the concrete application of new knowledge in everyday life using their consolidated expertise and know-how regarding the existing machinery.

The organizational strategy of Feralpi concerning Industry 4.0 has also fostered the initiative of "E-farmers." "E-farmers" are 12 young graduates, who were selected from different Italian universities and intentionally with different study backgrounds, not only technical ones. They were introduced in Feralpi with the goal to digitize internal processes, by proposing some disruptive projects in a competition context. The project which won is called ADAM (*Assistente Digitale Attivitá Manageriali*—Digital Assistant for Managerial Activities). It will be a digital assistant which, with the help of AI and machine learning, will simplify the interaction between the commercial departments of Feralpi and their customers, while reducing time of responses and operational repetitive activities. This initiative shows the goal of the management of integrating senior and junior employees: creating diversity in the staff is an opportunity to enrich innovative knowledge and competences of the company and to find new ways of solving problems.

Technologies of Industry 4.0 have improved the performance in circular economy initiatives. Very advanced sensors let capture deviations from the process and measure the level and the consistence of emissions. Process integration and machine learning lead to resource optimization and energy saving. Modeling and simulation permit to analyze the use of recycled materials, which are coming as output of the production process. Integrated platforms lead to exchange of information with established and potential business partners and to the creation of consortiums to explore new circular economy solutions. Feralpi's goal is to combine Industry 4.0 and sustainability aspects in order to exploit positive consequences of the circular economy.

Considering Industry 4.0 and sustainability, Feralpi is participating in an important EU project, started in 2015 and which will end in 2020, called FISSAC (Fostering Industrial Symbiosis for a Sustainable Resource Intensive Industry across the extended Construction Value Chain). This project is focused on the construction supply chain and has the goal to facilitate the reuse of industrial disposal material. The goal is to facilitate exchange of best practices and information, by developing a software platform in order to calculate the usability of disposal material in other local companies. The positive effect is to create an industrial symbiosis in the construction sector by the creation and use of eco-innovative construction products (new eco-cement and green concrete, innovative ceramic tiles, and rubber-wood plastic composites).[9]

The belief of Feralpi in sustainable growth is a concrete value, which is transmitted to the next generations, of employees, managers, and shareholders. It is the backbone of the technological development of the group (Interview Chairman

---

[9]Fissac (2019). https://fissacproject.eu/en/project/

Feralpi Holding, 2019a). In May 2019, Feralpi was awarded with the prize of "Best Performer in the Circular Economy" for the category Big Manufacturing Companies in Rome, in the presence of the Ministry for the Environment. This was another important acknowledgment that sustainability intertwined with advanced technological solutions is the right path in order to preserve competitive advantage.

4 Example of Application of Industry 4.0 in Other Steel Companies

Industry 4.0 is affecting the entire steel sector, with different industrial approaches. We will give some practical examples in the following section, and at the end, we will see which are the common patterns for all these industries, including Feralpi, concerning Industry 4.0 and technological and knowledge development.

Saarstahl AG is a German manufacturer of steel long products, with subsidiaries in Völklingen, Burbach, and Neunkirchen, counting 6590 employees and a crude steel production of 2.8 million tons (Saarstahl, 2019).

In 2015, Saarstahl started the project "iPRODICT," in collaboration with:

- Fraunhofer IAIS (Institute for Intelligent Analysis and Information Systems)
- Bundesministerium für Bildung und Forschung (German Federal Ministry of Education and Research)
- DFKI (Deutsches Forschungszentrum für Künstliche Intelligenz—German Research Centre for Artificial Intelligence)
- Blue Yonder GmbH
- PRC (Pattern Recognition Company)
- Software AG

This project is aimed to exploit IoT, especially sensors and video cameras at the continuous casting machine, and big data analysis, in order to recognize surface failure patterns, and connect them to the process causes. The model links and combines historical data to create patterns, evaluated and integrated in streaming production data.

This solution has enabled Saarstahl to avoid quality failure in semifinished products, which must be further processed by other subsidiaries, avoiding negative consequences on production efficiency and customer satisfaction. The challenge of the project was to decide the rules of the failure recognition models and to determine which production data were useful, correct, and existent and who in the production should have had access to the elaboration and interpretation of these data (Stahmer, 2016).

Thyssenkrupp is a diversified group with sales for almost €43 billion and 161,000 employees present in 78 countries (Thyssenkrupp Steel, 2019a, 2019b, 2019c). Thyssenkrupp Rasselstein is one of its subsidiaries, located in Andernach, with around 2400 employees, and producing ca. 1.5 million tons of packaging steel/year for international customers (Thyssenkrupp Steel, 2019a, 2019b, 2019c).

The company started a project of IoT and big data analysis, in collaboration with BIF (Research Center of the German Association of ironworks), in order to improve the surface analysis on steel coils. The project enabled the improvement of product tracking, evaluating data of thousand coils in a very fast time and in a very good visual quality and accuracy. It pushed for new organizational concept of quality control, allowing intervention in quality problems in real time (Brandenburger & Schirm n.d.).

Thyssenkrupp Hohenlimburg is another subsidiary located in Hagen, with around 900 employees. It produces hot-rolled special steel strip, which is primarily used in the automotive industry (Thyssenkrupp Steel, 2019a, 2019b, 2019c). The company strengthens its horizontal integration with customers via a smartphone application, the "tk HO app," available on all common platforms. Customers can access a private section, to view contracts data, like the status of an order, and to order new products. Via the smartphone camera, customers can read the barcode on the coil label and retrieve immediately all necessary information (Thyssenkrupp Steel, 2018).

Voestalpine Stahl GmbH in Linz is an important Austrian steel producer. Together with the Johannes Kepler University in Linz (JKU), they developed a special radar system to register the burden's behavior in high temperatures and dust conditions. Since no technology for this was available on the market, Voestalpine developed and patented this 3D radar technology. This unique, continuous 3D radar measurement can be used to monitor changes in the blast furnace in real time. The technology used is comparable to the technology for radar distance measurements used in the automotive industry to avoid collisions (Voestalpine, 2018).

Salzgitter AG, German producer, is one of the largest steel producers in Europe, with a crude steel production of 7 million tons, a turnover of 9 billion euro, and 23,139 employees (2017) (Salzgitter AG, 2019).

During the fair Hannover Messe 2017, Salzgitter AG presented "HelmetGlass," created by its technological subsidiary GESIS Gesellschaft für Informationssysteme mbH. "HelmetGlass" gives workers in production all the necessary information concerning the working environment, thanks to a Bluetooth interface. If there is a danger, it is displayed directly in the user's virtual view. Moreover, these glasses display the list of tasks to be fulfilled by the operator or machine conditions. During maintenance operation, the operator is connected via audio and video to the remote central service center, which guides him/her through the different activities (Stahlblog.de, 2018).

Tata Steel Europe is one of the largest steel producer. It belongs to the Tata Group, which includes 29 publicly listed enterprises, with a total market capitalization of $103.51 billion and 660,800 employees worldwide (2017) (Tata Steel, 2019a, 2019b, 2019c, 2019d).

Tata Steel Europe counts 21,000 employees, €7.9 billion sales, and 10 million tons delivered steel products (to March 2018) (Tata Steel, 2019a, 2019b, 2019c, 2019d).

In 2015, Tata launched the Innovation Portal, in order to be fertilized by ideas by other sectors. Up to now, Tata received 380 proposals from different companies and research centers, including ideas about to extend the life and improve the quality of products, to reduce water usage, and to employ laser technologies to measure the shape of products in-line and also the use of high-speed camera technologies developed for the postal industry to monitor the quality of our strip products in-line (Tata Steel, 2019a, 2019b, 2019c, 2019d). Tata integrated this program with Tata Innoverse, a portal where Tata is seasonally posting concrete challenges, rewarded with money and with a prize in an annual meeting with the management. With the program Start-O-Sphere, Tata also wants to partner with startups and technologists (Tata Steel, 2019a, 2019b, 2019c, 2019d).

With annual production of approximately 93 million tons of crude steel (47% manufactured in Europe), $US 69 billion of turnover, and 199,000 employees across 60 countries, ArcelorMittal is the world's leading steel and mining company (Arcelor Mittal Fact Book, 2017).

ArcelorMittal digitized its supply chain with the introduction of "SteelUser," a platform used by 89% of its customers, in order to access all information about orders at any time. The platform is supported by the application "Track & Trace," allowing a mobile shipment tracking, too, and by the additional Internet platform "SteelAdvisor," which supports customers in selecting the right steel product for a job.

All these tools helped enhancing customer relation and reduced errors and, above all, administrative costs (Stahlblog.de, 2018).

Vallourec is a multinational enterprise active in 20 countries and counting 19,000 employees. It produces tubular solutions for the energy market. In 2016, it had sales of €2.9 billion (Vallourec, 2019a). Vallourec started to offer their customers more than a platform. They offer a complete service for the whole supply chain: design of products, installation, sourcing, and asset management, by inserting sensors in final products in order to monitor them in their service life (Vallourec Smart, 2019b).

NSSMC (Nippon Steel and Sumitomo Metal Corporation) is the leading Japanese corporation with 13 steelworks in Japan. It is specialized in three business fields as key strategic areas: high-grade steel products for automobiles, resources and energy, and civil engineering, construction, and railways (NSSMC, 2019). It possesses three major R&D research centers, and seven laboratories at steelworks, all in Japan, with focus on steelmaking, engineering, chemicals, new materials, and system solutions. The NSSMC group employs approximately 84,000 persons (Nippon Steel and Sumitomo Metal Report, 2018). Beyond their core business, they developed a specialized System Solution segment, offering customers complete solutions in the configuration, operation, and maintenance of IT systems. Customers are supported in the employ of AI, machine learning, and IoT. The system solutions segment recorded net sales of $1.9 billion (Nippon Steel and Sumitomo Metal Report, 2018).

The trend to assist customers in their whole purchasing experience is diffusing in the steel sector.

A further example is Klöckner & CO. Klöckner & CO is one of the largest producer-independent distributors of steel and metal products worldwide. With around 170 locations in 13 countries, the group supplies ca. 120,000 customers. Klöckner & CO's target is to fully digitalize its supply and service chain as well as to expand the independent open industry platform XOM Materials to become the dominant vertical platform for the steel and metal industry (Klöckner, 2019).

The digitization strategy of Klöckner consists in those pillars:

1. Digitalized platforms: creation of industrial platforms to disrupt the traditional steel and metal trading and to exploit the opportunities of the e-commerce.
2. Higher value-added business: digitization will give space to focus on businesses with higher margins and to automate repetitive and less profitable activities.
3. Higher efficiency: Klöckner launched the internal program "VC^2—Value Creation at the Core." It is an accelerator of operational initiatives aimed at achieving internal excellence and profitable growth and to support a cultural transformation across all parts of our group (Klöckner, 2019).

In 2018, Klöckner started the first version of their open industry platform, XOM Materials. In contrast to the Klöckner & Co online shops, the industry platform is a completely independent digital marketplace for steel, metal, and industrial products and is open to direct competitors. XOM Materials is controlled by external investors, which guarantee the independence from Klöckner (Klöckner, 2019).

Moreover, Klöckner & Co founded two spin-off companies: Klöckner.i, based in Berlin, taking care of all projects and initiatives relating to Klöckner & Co's digitalization, and Klöckner Ventures, a second spin-off company, investing in B2B disruptive digital business models, using IOT solutions, and collaborating with startups and venture capital funds (Klöckner, 2019). Klöckner also invested in the development of AI, in alliance with Aera and Arago, two leading providers of AI.

The clear strategy of Klöckner to become a digital disruptor in the steel sector is not only challenged by its competitors but also by some giants of the e-commerce. Nowadays, it is already possible to purchase industrial products on platforms like Amazon Business and Alibaba.

The trend in the industrial sector is to experiment collaborative platforms and ecosystems in order to exploit reciprocal advantage from the completion and synchronization of their own database.

Another example in this sense is the project PRODISYS. The project is financed by the German Federal Ministry of Education and Research (Bundesministerium für Bildung und Forschung—BMBF) with a total volume of €3 million and runs to June 2020. The goal is to create a prototype of an industrial platform ecosystem, with connected and combinable services for improved planning and coordination of production and connected areas. In addition to the conceptual reorientation of service systems for bundling and organizing services, the integration of system, order, and production data is addressed. Both service integration and data integration require appropriate models and digital representations. Implementation partners also identify productivity factors and analyze the value-added flows based on the resulting models. Such system could enable the networking of services and processes among the partners involved, with important synergy effects, with a continuous data exchange (Lehmann, 2018). Research partners are:

- Fortiss GmbH: coordinator of the consortium, it is the research institute of the Free State of Bavaria for software-intensive systems and service (Fortiss, 2019).
- Friedrich-Alexander-University Erlangen-Nürnberg: with the chairs of business informatics, Manufacturing Automation, and Production Systematics (FAU Nürnberg, 2019).
- Center for Leading Innovation & Cooperation (CLIC).

Application partners are:

- Audi AG
- Continental Automotive GmbH
- Crossbar.io: an open source networking platform for distributed and microservice applications
- SAP SE
- XENON Automatisierungstechnik GmbH: it develops and builds assembly lines and inspection lines to automate the manufacture of mechatronic components

Industrial platforms are also affecting the way to manage and control circular economy processes. Sfridoo is a startup and an example of a market platform for B2B purchasing and selling of scrap materials, secondary raw materials, and by-products.

Beyond the publishing platform, it offers private messaging for communication between companies and geolocalization about scrap materials. It has more than 140 registered companies and it offers additional digital tools to register, monitor, and sell current and potential certified by-products (Sfridoo, 2019).

Industry 4.0 is pushing steel industry not only to become smart factories but also to contribute to create smart industrial environments and smart cities. In the following table, we tried to summarize the most used tools of Industry 4.0 and their principal areas of application in the steel sector (Table 1).

POSCO, positioned in South Korea, is one of the biggest steel producer in the world, with 37,000 employees and 41.6 million tons of steel (Posco, 2019). They created a smart factory concept by developing a platform collecting big data, through the use of numerous sensors, interpreted by AI processes (Ferneyhough, 2018). At the same time, Posco is investing money in R&D to analyze which steel grades will be used in the smart cities of the future. Construction materials will be recycled or used in smaller quantities; buildings will be restored or rebuilt rather than being constructed from scratch; high-strength steel materials will be increasingly used for supertall buildings and super-long-span bridges in megacities; and steel content per unit of construction investment is expected to decline (Hoon-sang, 2017).

5 Conclusions

Industry 4.0 is affecting the steel sector in three different ways: (1) from an industrial, (2) from an organizational, and (3) from a market perspective.

The industrial perspective is the most natural way for steel industries to introduce Industry 4.0 technologies. Steel industries have always invested in innovative

Table 1 Industry 4.0: perspectives from the steel sector

Goals	Tools	Area of application
1. Safety and comfort at the workplace	Robotized labeling and sampling systems in the steel plant and in the rolling mill	Advanced manufacturing solutions
2. Optimization of resources/ sustainable production process	Use of simulation programs based on KPIs. Use of advanced sensors and automation integrating and adapting the production process to the most performative outcome	Simulations, IoT, machine learning
3. Predictive maintenance	Use of sophisticated sensors and self adaptive systems retrieving information of possible breaks and breakdowns in machines	IoT, machine learning, AI
4. Predictive quality/higher product quality	Recognition of surface failure patterns on final products with the help of sensors and video cameras	IoT, big data analytics
5. More effective training to employees; enhanced safety	Use of visors and tablets visualizing information for tasks and maintenance directly near machines. Information about important safety issues and alarms	Augmented reality
6. Personalized service to customers	Use of web portals and application to communicate directly with customers and make the buying experience easier and more comfortable	Cloud, horizontal integration
7. New services to customers	Assistance to customers in the digital transformation of their company	Horizontal integration
8. Integration in the supply chain	Creation of industrial portals connecting data and machines of different stakeholders (supplier, producers, customers)	Horizontal and vertical integration
9. Open innovation	Openness to innovation experience of other sectors, making alliances with startups and other disrupting emerging companies	Horizontal integration
10. Shared business models	Collaboration with companies of adjacent or different sectors to employ by-products and final products innovatively	Horizontal integration

technologies in order to produce at the best costs and in the most efficient way. Technology innovation goes together with sustainable growth, as we clearly analyze in the case of Feralpi. Industry 4.0 and sustainability are strongly correlated in the steel business. With the legal constraints in the EU very stringent concerning emissions and environmental impact, steel industries are obliged to employ the Best Available Technologies in order to have less impact on the local community. Digital technologies also give the possibility to steel producers to have a more transparent and interactive communication with their stakeholders, which ignites a sense of belonging and respect toward the best practices fulfilled, often with intense economic effort, in the company.

Industry 4.0 has significant impact on the organization of steel industries. Managers have to incentivize and support the staff in the acquisition of new competences and in the process of cultural change, through the acceptance and assimilation of new ways of working. Industry 4.0 technologies are partially influencing the technical assets of the company, while they are hugely affecting the knowledge of human resources. The employ of Industry 4.0 technologies implies a considerable organizational effort of rethinking and redesigning their factory not for the short-middle term but for a long-term perspective, as if the company could always adapt itself to the changing environment.

Finally, Industry 4.0 gives steel industries a new market perspective, intended as new business opportunities and new fruitful cooperation in and outside the own supply chain. As we could observe, the trend is to be more integrated to the business partners to offer new services or more qualitative products, but at the same time to collect new solutions from other perspectives, like from other industrial or even non-industrial sector, in a logic of open innovation.

It will be certainly worth deepening the research on Industry 4.0 in the steel sector, by looking for the concrete application beyond the trend and, at the same time, by comparing best practices from other industrial businesses. Beyond the economic profitability, sustainability could be an additional interesting perspective to analyze the advantages and the challenges of Industry 4.0 in such a traditional but fundamental sector, as steel business is.

References

ArcelorMittal Fact Book. (2017). ArcelorMittal, Luxembourg. https://corporate.arcelormittal.com/~/media/Files/A/ArcelorMittal/investors/fact-book/2017/factbook-2017.pdf

ASI, Agency for Strategic Initiatives. (2019). *National technology initiative*. https://asi.ru/eng/nti/

Berg, A. (2019, April). Industrie 4.0 – Jetzt mit KI. *Bitkom*. https://www.bitkom.org/sites/default/files/2019-04/bitkom-pressekonferenz_industrie_4.0_01_04_2019_prasentation_0.pdf

BMBF: Bundesministerium für Bildung und Forschung. (2018). https://www.softwaresysteme.pt-dlr.de/de/embedded-cyberphysical-systems.php

Boschi, F., De Carolis, A., & Taisch, M. (2017, January). Cosa sono i CPS e in che modo abiliteranno/faciliteranno la trasformazione manifatturiera. *Industria Italiana*, Torino. https://www.industriaitaliana.it/nel-cuore-dell-industry-4-0-i-cyber-physical-systems/

Brandenburger, J., & Schirm, C. (n.d.) Presentation. Big-Data-Implementierung zur Qualitätsbewertung von Flachstahlprodukten [PowerPoint slides]. VDEh-Betriebsforschungsinstitut.

Brunelli, J., Lukic, V., Milon, T., & Tantardini, M. (2017). *Five lessons from the frontlines of industry 4.0*. Boston Consulting Group. https://www.bcg.com/it-it/publications/2017/industry-4.0-lean-manufacturing-five-lessons-frontlines.aspx

Bundesregierung. (2015, February). *Merkel: Standards für "Industrie 4.0" entwickeln*. https://www.bundesregierung.de/breg-de/aktuelles/merkel-standards-fuer-industrie-4-0-entwickeln-746752

CLIC (Center for Leading Innovation and Cooperation). (2019). http://clicresearch.org/about-us/

Di Maria, E., Valentina De Marchi, Silvia Blasi (2018). *L'economia circolare nelle imprese italiane e il contributo di Industria 4.0*. Dipartimento di Scienze Economiche e Aziendali "Marco Fanno". Legambiente.

Eloot, K., Huang, A. & Lehnich, M. (2013, June). *A new era for manufacturing in China*. McKinsey & Company. http://www.mckinsey.com/business-functions/operations/our-insights/a-new-era-for-manufacturing-in-china

Eurofer. (2019). *European steel in figures*. The European Steel Association. http://www.eurofer.org/News%26Events/News/20190703%20European%20Steel%20in%20Figures%202019.fhtml

European Parliament. (2016). Directorate General for Internal Policies. Policy Department. Economic and Scientific Policy. Industry 4.0. Study for the ITRE Committee. http://www.europarl.europa.eu/RegData/etudes/STUD/2016/570007/IPOL_STU(2016)570007_EN.pdf

FAU Nürnberg. (2019). *Friedrich-Alexander Universität Erlangen-Nürnberg*. https://www.fau.de/

Feralpi Group Sustainability Report. (2018). *Bilancio di sostenibilità. Esercizio 2018*. Lonato del Garda: Feralpi Holding SpA.

Feralpi Holding S.p.A. (2019). http://www.feralpigroup.com/en/innovazione-en/

Ferneyhough, G. (2018). *Steel rises to the challenges of industry 4.0*. World Steel Association. https://stories.worldsteel.org/innovation/steel-rises-challenges-industry-4-0/

Fissac. (2019). https://fissacproject.eu/en/project/

Fortiss. (2019). https://www.fortiss.org/en/about-us/institute/overview/

GRI. (2019). *Global reporting initiative*. https://www.globalreporting.org/Pages/default.aspx

Hauser, H. (2014). Review of the Catapult network. Recommendations on the future shape, scope and ambition of the program. Crown copyright, London.

Hoon-sang, K. (2017). *Future cities and changes in steel materials*. POSCO Research Institute. https://www.posri.re.kr/files/file_pdf/59/334/6793/59_334_6793_file_pdf_1499150152.pdf

Industrial Internet Consortium. (2019). https://www.iiconsortium.org/about-us.htm

Interview. (2019a). Chairman Feralpi Holding SpA.

Interview. (2019b). CSR Manager Feralpi Holding SpA.

Interview. (2019c). Managing Director Feralpi Holding SpA.

Interview. (2019d). Responsible for Environment. Feralpi Siderurgica SpA.

Interview. (2019e). Responsible of Industry 4.0 Feralpi Holding SpA.

IVI: Industrial Value Chain Initiative. (2019). https://iv-i.org/wp/en/about-us/whatsivi/

IW Consult. (2017). *Potentiale des digitalen Wertschöpfungsnetztes Stahl. Die Rolle der Stahlindustrie als Enabler der Digitalisierung der deutschen Wirtschaft*. Studie im Auftrag der Wirtschaftsvereinigung Stahl. IW Consult, Köln.

Janjua, R. (2018, June). *Blog: Steel digital strategy guided by standardization*. https://www.worldsteel.org/media-centre/blog/2018/Steel-digital-strategy-guided-by-standardisation.html

Kagermann, H., Anderl, R., Gausemeier, J., Schuh, G., & Wahlster, W. (2016). *Industrie 4.0 im globalen Kontext. Strategien der Zusammenarbeit mit internationalen Partnern (acatech STUDIE)*. München: Herbert Utz Verlag.

Kepmf, D. (2018, May). *Klartext "Den digitalen Wandel unterstützen"*. 1/2018. Stahl Das Magazin, Wirtschaftsvereinigung Stahl. https://issuu.com/stahlonline/docs/rz_stahl_magazin_06_20180302_web/24

Kerkhoff, H. J. (2019, March). Stahlindustrie in Deutschland 2019 – Zwischen globalen Herausforderungen und regionalen Perspektiven. https://www.stahl-blog.de/index.php/stahlindustrie-in-deutschland-2019-zwischen-globalen-herausforderungen-und-regionalen-perspektiven/

Klitou, D., et al. (2017a, January). *Digital transformation monitor*. France Industrie du Futur, European Union.

Klitou, D., et al. (2017b, January). *Digital transformation monitor. Germany: Industry 4.0*. Brussels: European Union.

Klitou, D. et al. (2017c, January). *Digital transformation monitor. Spain. Industria Conectada*. Brussels: European Union

Klitou, D., et al. (2017d, May). *Digital transformation monitor. Key lessons from national industry 4.0 policy initiatives in Europe*. Brussels: European Union

Klitou, D. et al. (2017e, May). *Digital transformation monitor. Country: Portugal "Indústria 4.O"*. Brussels: European Union

Klitou, D., et al. (2017f, June) *Digital transformation monitor. Denmark: Manufacturing Academy of Denmark (MADE)*. Brussels: European Union

Klöckner. (2019). *"Klöckner & CO 2022" – Our strategy.* http://www.kloeckner.com/en/strategy.html?langSwitched=1

Lee, X. (2015, September). *Made in China 2025: A new era for Chinese manufacturing.* CKGSB Knowledge. China-focused leadership and business analysis. http://knowledge.ckgsb.edu.cn/2015/09/02/technology/made-in-china-2025-a-new-era-for-chinese-manufacturing/

Legambiente. (2019). https://www.legambiente.it/cosa-facciamo/

Lehmann, C. (2018). *Industrie 4.0: Moving from "what" to "how".* Leipzig: CLIC: Center for Leading Innovation and Cooperation.

LMD: Laboratorio Manifattura Digitale (2019). https://economia.unipd.it/LMD/laboratorio-manifattura-digitale

Make in India. (2019). About us. http://www.makeinindia.com/about

Maniyka, J. (2011, May). *Big data: The next frontier for innovation, competition, and productivity.* Washington, DC: McKinsey Global Institute

Marr, B. (2018, September). *What is industry 4.0? Here's a super easy explanation for anyone.* Forbes Media LLC

MISE: Ministero dello Sviluppo Economico. (2018) *Piano Nazionale Industria 4.0. Investimenti, produttività e innovazione.* https://www.mise.gov.it/images/stories/documenti/investimenti_impresa_40_ita.pdf

Nippon Steel & Sumitomo Metal Report (2018, March). NSSMC report 2018. https://www.nipponsteel.com/en/ir/library/pdf/nssmc_en_ar_2017_all_a4.pdf

NSSMC. Nippon Steel & Sumitomo Metal Corporation. (2019). https://www.nipponsteel.com/

OECD: Organization for Economic Co-operation and Development (2019, March). *OECD Steel Committee concerned about excess capacity in the steel sector.* https://www.oecd.org/industry/oecd-steel-committee-concerned-about-excess-capacity-in-steel-sector.htm

Pareekh, J., & Tanmoy, M. (2017, April). *Blueprint guide: Industry 4.0 services.* Excerpt for Accenture. HfS. https://www.accenture.com/t20180625t094010z__w__/us-en/_acnmedia/pdf-52/accenture-industry-4-excerpt-for-accenture-report.pdf

Peterson, B. (2018, August). La blockchain spiegata bene. Ecco la nuova tecnologia informatica che potrebbe essere dirompente quanto internet. *Business Insider.* https://it.businessinsider.com/la-blockchain-spiegata-bene-ecco-la-nuova-tecnologia-informatica-che-potrebbe-essere-dirompente-quanto-internet/

Pflaum Dr, A., et al., (2014) Industrie 4.0 und CPS – Bedarfe und Lösungen aus Sicht des Mittelstands. bayme Bayerischer Unternehmensverband Metall und Elektro e. V. and vbm Verband der Bayerischen Metall- und Elektro-Industrie e. V., München

Posco. (2019). http://www.posco.com/homepage/docs/eng6/jsp/s91a0000001i.jsp

Rouse, M. (2018, August). AI: Artificial intelligence. *TechTarget.* https://searchenterpriseai.techtarget.com/definition/AI-Artificial-Intelligence

Saarstahl. (2019). http://www.saarstahl.com/sag/de/konzern/medien/presse/iprodict-bringt-big-data-und-big-steel-zusammen-67161.shtml

Salzgitter, A. G. (2019). https://www.salzgitter-ag.com/en/company/key-data.html

Santoro, F. (2019, June). *Europa-Cina. La sfida industriale di "Made in China 2025".* https://www.geopolitica.info/made-in-china-2025/

Sargent, J. (2013). *Federal research and development funding: FY2013.* Washington, DC: Congressional Research Service.

Saville, J. P. (2019). *Henry Bessemer Britannica.* https://www.britannica.com/biography/Henry-Bessemer

Schumpp, F., Birenbaum, C., & Schneider, M. (2019, February). *Digitalisierung im Branchenfokus Stahl- und Metallhandel.* Frauernhofer IPA

Sfridoo. (2019). https://www.sfridoo.com/

Son, J. (2018). *Korea manufacturing technology smart factory.* https://www.export.gov/article?id=Korea-Manufacturing-Technology-Smart-Factory

Stahlblog.de (2018). https://www.stahl-blog.de/index.php/stahl-4-0-digitale-loesungen-in-der-stahlindustrie/

Stahmer, B. (2016). Presentation. Vorstellung des Verbund-forschungsprojekts "iPRODICT". Intelligent process prediction based on big data analytics. [PowerPoint slides]. Düsseldorf, Saarstahl.

Tata Steel. (2019a). *Innoverse*. https://www.tatainnoverse.com/startosphere

Tata Steel. (2019b). *Open innovation*. https://www.tatasteeleurope.com/en/innovation/open%E2%80%93innovation

Tata Steel. (2019c). *Profile*. https://www.tatasteel.com/corporate/our-organisation/tata-group-profile/

Tata Steel. (2019d). *Tata Steel in Europe*. https://www.tatasteeleurope.com/static_files/Downloads/Corporate/Tata%20Steel%20In%20Europe%20Factsheet%20ENG.pdf

Thyssenkrupp Steel. (2018). https://www.thyssenkrupp-steel.com/en/newsroom/press-releases/press-release-53504.html

Thyssenkrupp Steel. (2019a). *Figures*. https://www.thyssenkrupp.com/en/company/key-figures

Thyssenkrupp Steel. (2019b). *Rasselstein*. https://www.thyssenkrupp-steel.com/en/company/locations/thyssenkrupp-rasselstein-gmbh/

Thyssenkrupp Steel. (2019c). *Hohenlinmburg*. https://www.thyssenkrupp-steel.com/en/company/locations/hoesch-hohenlimburg-gmbh/

Vallourec. (2019a). http://www.vallourec.com/EN/Pages/default.aspx

Vallourec Smart. (2019b). *Vallourec Smart 2019*. http://www.vallourec.com/EN/group/MEDIA/Publications/Documents/LEAFLET_VALLOUREC_SMART_EN.pdf

Voestalpine. (2018). *Voestalpine develops 3 radar technology for blast furnaces*. https://www.voestalpine.com/blog/it/innovation-en/voestalpine-develops-3d-radar-technology-for-blast-furnaces/

World Steel Association. (2018). *Sustainable steel. Indicators 2018 and industry initiatives*. https://www.worldsteel.org/en/dam/jcr:ee94a0b6-48d7-4110-b16e-e78db2769d8c/Sustainability%2520Indicators%25202018.pdf

World Steel Association (2019). *World steel in figures*. https://www.worldsteel.org/en/dam/jcr:96d7a585-e6b2-4d63-b943-4cd9ab621a91/World%2520Steel%2520in%2520Figures%25202019.pdf

World Steel Organization. (2019). *Steel facts. The uses of steel*. https://www.worldsteel.org/about-steel/steel-facts.html

CPSIA information can be obtained
at www.ICGtesting.com
Printed in the USA
LVHW051248140621
690169LV00001B/41